THE SKY IS NOT THE LIMIT

ALSO BY NEIL deGRASSE TYSON:

MERLIN'S TOUR OF THE UNIVERSE

UNIVERSE DOWN TO EARTH

JUST VISITING THIS PLANET

ONE UNIVERSE: AT HOME IN THE COSMOS
(with Charles Liu and Robert Irion)

COSMIC HORIZONS
(with Steven Soter, Editors)

ORIGINS: FOURTEEN BILLION YEARS OF COSMIC EVOLUTION
(with Donald Goldsmith)

THE SKY IS NOT THE LIMIT

ADVENTURES OF
AN URBAN
ASTROPHYSICIST

NEIL DeGRASSE TYSON

 Prometheus Books

59 John Glenn Drive
Amherst, New York 14228-2119

Published 2004 by Prometheus Books

Inquiries should be addressed to
Prometheus Books
59 John Glenn Drive
Amherst, New York 14228–2119
VOICE: 716–691–0133, ext. 210
FAX: 716–691–0137
WWW.PROMETHEUSBOOKS.COM

13 12 11   6 5 4 3

Library of Congress Cataloging-in-Publication Data

Tyson, Neil deGrasse.
    The sky is not the limit : adventures of an urban astrophysicist / Neil deGrasse Tyson.
      p. cm.
    Includes bibliographical references and index.
    ISBN 978–1–59102–188–9 (pbk. : alk. paper)
    1. Tyson, Neil deGrasse. 2. Astrophysicists—United States—Biography. I. Title.
QB460.72.T97A3 2004
523.01'092—dc22

2004003989

Printed in the United States of America on acid-free paper

For Miranda and Travis

In the hope that the stars for which they reach
sit higher and brighter than any I have known.

CONTENTS

PREFACE

Memoirs are both easy and hard to write. The material was all there—the details of my scientific journeys were either stored in my memory or retrieved from rather extensive records I have kept from early childhood. I could easily assemble the interesting parts and write about them.

But I am neither a movie star nor a sports celebrity, nor am I an important political figure. These various professions make fertile lifestyles for the memoir format and typically attract wide readership. In my case, however, I am just a scientist—an astrophysicist—who has tried to bring the universe down to Earth for everybody who wanted to have a look. And, in what I consider a privilege, I have also tried to elevate public literacy in science.

Why then might you be interested in my story?

In these pages, I share what I believe to be amusing and playful moments of my life in the cosmos. But I also share the segments of my life's path that got me here, which, for the most part, tacked against the winds of society. The paths include fond memories of mentors—some of whom were ordinary people doing extraordinary things while others were extraordinary people doing ordinary things—and the retelling of traumatic moments where my will, my life's goals, and my sense of identity were tested to their limits.

Regardless of what you may seek in my or anyone else's memoir, I can promise you that *The Sky Is Not the Limit* will bring you closer to the universe of ambition; and, as is my life's commitment, bring you closer to the universe itself.

Neil deGrasse Tyson
New York City
April 2004

ACKNOWLEDGMENTS

Of the hundreds of people who have been there for me over the years, I single out my wife, Alice; my father, Cyril; my mother, Sunchita; my brother, Stephen Sr.; and my sister, Lynn. Through their continual love and support, they have collectively supplied an emotional and intellectual buoyancy to my life's journeys. By way of their advice, wisdom, and guidance, I have cleared life's hurdles and survived life's challenges. For this, I owe them more than I have the capacity to express.

Betsy Lerner, my editor, has supported my writings and has observed my career since my later years in graduate school. She alone encouraged me to write this memoir and I am thankful for her persistence, in spite of my stubborn apprehensions about undertaking such a project.

Portions of chapters 5 and 7 are adapted from essays that touched upon my life and were originally written for *Natural History* magazine under my column titled "Universe."

Beyond the judgments of others
Rising high above the sky
Lies the power of ambition

—Neil deGrasse Tyson

INTRODUCTION

At my high school graduating class's twenty-year reunion, during the obligatory assessments of how well time had treated us all, I won "the coolest job" contest in a straw poll of all those attending. That particular graduating class from New York City's Bronx High School of Science was not unusual. It had produced the typical ensemble of scientists, medical doctors, lawyers, and the like. But I was the only one who had keys to the candy store, so who was I to argue the honor? As an astrophysicist and as the director of New York City's celebrated Hayden Planetarium, I get to decode the nature of the universe and create journeys through it for all the public to see. Like countless people before us, we had all visited the Hayden Planetarium as kids, if not with our parents then with our grade-school classmates and teacher. For school trips, we had all been to the Bronx Zoo, the New York Botanical Gardens, the Cloisters, and other cultural offerings of the city. But none were as magical, nor, of course, as otherworldly, as the Hayden Planetarium, implanting memories like none other from our childhood.

What was largely unknown to my classmates, however, was the unorthodox profile I carried into this coolest of jobs. Although everyone leads a unique life, certain categories of experience can be justifiably generalized: My tenure as a nerdy kid—complete with winnings in the science fair, membership in the physics club, and high scores in mathematics—greatly resembles all that you may have stereotyped for the community of nerds. My time as an athlete—as captain of my high school's wrestling team and as a varsity competitor in college—was not fundamentally different from that of any other athlete. My interest in the universe—carrying me to a PhD in astrophysics—led me down paths shared by many of my colleagues. And my life as a black man in America—getting stopped for no reason by the police or being trailed by security guards in department stores—is hardly different from that of other black men among my contemporaries. But when you combine all these ingredients into one package, my experiences offer what may be an uncommon portal through which to view life, society, and the universe.

More a rumination than a memoir, I wrote *The Sky Is Not the Limit* in a way that reveals to the reader how scientists view the world—how I view the world. I want every generation of stargazers—whether they sit on a tenement roof or an Appalachian mountain—to have a fresh lens with which to see the universe and reach for their own star.

I.
NIGHT VISION

The Early Years

It was a dark and starry night. The sixty-five-degree air was calm. Visibility was unlimited. Too numerous to count, the stars of the autumn sky, and the constellations they trace, rose slowly in the East while the waxing crescent moon descended into the western horizon. High in the northern sky were the Big Dipper and Little Dipper, just where they were supposed to be, just as they were supposed to appear. The planets Jupiter and Saturn were also high in the sky. One of the stars, I don't remember which, seemed to fall from the sky in a flash of light toward the horizon. No, I was mistaken. A meteor had just vaporized, leaving a glowing trail through the atmosphere. I was told there would be no clouds that night, but I saw one—long and skinny—meandering across the sky from horizon to horizon. Once again, I was mistaken. What I traced was not a cloud but the Milky Way—with its varying bright and dark patches giving the appearance of structure and the illusion of depth. I had never seen the Milky Way with such clarity and majesty as that night, that dark and starry night. Forty-five minutes of my suspended disbelief swiftly passed when the house lights gently undimmed back to full strength under the dome of the Hayden Planetarium sky theater in the middle of Manhattan.

That was the night. I had been called. The study of the universe would be my career, and no force on Earth would stop me. I was just nine years old, but I now had an answer for that perennially annoying question adults ask children, "What do you want to be when you grow up?" Although I could barely pronounce the word, I would thenceforth reply, "I want to be an astrophysicist."

For years to follow, one question lingered within me: Was that indelible planetarium sky an accurate portrayal of the real celestial sphere? Or was it a fantasy? Or worse, a hoax? Surely there were too many stars. I had proof because I had seen the night sky from the Bronx—from the tarred rooftop of my apartment. Built upon one of the highest hills of the borough, my building was the first of a set of three, prophetically named the "Skyview" apartments, positioned one after another, north-south along the Hudson River.

In the northernmost of these buildings lived Phillip Branford, a close friend and classmate from fourth grade. He lived in a single-parent home with an older brother and sister, both of whom had active social agendas. The father, who retained custody of the three kids after the divorce, worked long hours. During my occasional visits Mr. Branford was only rarely at home. Phillip, instead, spent tons of time over at my place, especially on the weekends, never failing to complain how strict my parents were, with limits set on my playtime that had no counterpart in his household. I'm guessing his father assumed that the stability of my two-parent home and upbringing would add some structure and discipline to Phillip's life. While this may have been true, I am certain that his influence on me was far greater. With or without structure to his home life, Phillip was smarter than me in practically every way one would gauge such a thing. He taught me to play chess, poker, pinochle, Risk, and Monopoly. He introduced me to brain-teaser books, which, if you are unfamiliar with the genre, are books that resemble collections of those dreaded word problems from your high school math class. Well-written brain teasers, however, contain clever O. Henry–like plot twists in their answers that trick you with their simplicity. My favorite was this: Start with four ants, one on each corner of a square board that measures twelve inches on each side. Each ant decides to walk at the same speed directly toward the ant to its right. By the time all four ants meet in the middle of the table, how far has each one traveled? (Answer: twelve inches.)

Or, start with a brand-new, unshuffled deck of cards. Or simply sort them by suit and sequence them by number (typical of their arrangement when first purchased). Cut the deck, just as one might do before a card game, but do it one hundred consecutive times. What are the chances that all fifty-two cards will still be sorted by suit *and* ordered by number? (Answer: 100 percent.)

I loved teasers that involved math: Counting one number per second, how long would it take to reach a trillion? (Answer: 31,710 years.) And more entertaining problems like this: How many people must you collect into a room before you have a better-than-even chance that two of them would have the same birthday? (Answer: twenty-four.)

The more we played, the more stretched and sharpened my eleven-year-old brain became.

Phillip's most important contribution to my life's path, however, was introducing me to binoculars. I had used them before—primarily to view sporting events and to look in other people's windows. City dwellers don't normally have more than these two uses for binoculars. In this particular case, Phillip instead encouraged me to look up; to look beyond the streetlights, beyond the buildings, beyond the clouds, and out toward the Moon and stars of the night sky.

Nothing I can write will capture the acute cosmic imprinting from my first view from the Bronx of the waxing crescent moon across the Hudson River, high above the Palisades of New Jersey. Through those 7×35 binoculars, the Moon was not just bigger, it was better. The coal-dark shadows sharply revealed the Moon's surface to be three dimensional—a rich moonscape of mountains and valleys and craters and hills and plains. The Moon was no longer just a thing in the sky—it was another world. And if simple binoculars could transform the Moon, imagine what mountaintop telescopes could do with the rest of the universe.

I would later learn that Galileo was the first person in the world to look up with a good enough telescope to see what no one before him had ever dreamed: structure on the lunar surface, spots on the Sun, Venus going through phases (just like the Moon), Saturn and its rings, Jupiter orbited by moons of its own, and stars composing the faint glow of the Milky Way. When I, too, first saw these images I communed with Galileo across time and space. My cosmic discoveries, although old news for society, were as fresh for me in the Bronx, New York, as they must have been for Galileo in Florence, Italy, four centuries ago. Not only that, Galileo's "observatory" was his rooftop and his windowsill. So was mine.

These episodes of enlightenment, and others to follow, all occurred outside of formal educational structures and programming—they happened outside the classroom. Teachers, however, especially elementary school teachers, know little of a student's extracurricular behavior or interests. So when I compare my life's trajectories inside the classroom with those outside the classroom, a serious disconnect prevails. Hardly any of my schoolteachers— none from grades one through six, nor any from grades eight through twelve—would have predicted my current station in life. In elementary school, nobody ever said, "Neil will go far," or "Neil shows great potential," or "We expect great things from Neil." Most teachers probably assumed that I would one day be pumping gas somewhere, given that I was no one's model student. My grades and classroom performance were not high enough to distinguish me, and the teachers judged my social energy as disruptive, including one who commented to my mother during a parent-teacher conference, "Your son laughs too loud."

The best schoolteachers first evaluate the entire talent set repre-

sented within each student, and then they help explore career paths
that align with the student's interests. The worst teachers simply
issue statements that pass judgment on your behavior in their
attempt to homogenize it with the rest of the class.

For the comments section of my third-grade report card, Mrs.
O'Connell complained, "Neil should cultivate a more serious attitude toward his school work." In the same section of my fourth-
grade report card, Miss Taylor logged no comments for any of the
year's three sessions. Come fifth grade, nothing seemed to improve.
Mr. Goldman said of my behavior, while simultaneously playing
the roles of good cop and bad cop, "Neil is a good leader. He shows
that he respects the rights, dignity and feelings of others. He is
somewhat lax about completing his work, compositions, notebook,
etc. He needs to be encouraged and prodded." I suppose he noticed
that my social energy was actually amounting to something he saw
worth complimenting. But once again, the traditional measures of
school success were not forthcoming. In my sixth-grade report
card's first session, Mrs. Kreindler commented in roundhand cur-
sive, sternly, almost maternally, "Less social involvement and more
academic diligence is in order!"

Mrs. Kreindler was a tall, assertive woman with a keen sense of
academic discipline. She was probably the smartest teacher in my
elementary school, P.S. 81 on Riverdale Avenue in the Bronx. (In
spite of how often reporters feel compelled to place my roots in the
South Bronx, or in some other Bronx neighborhood where gangs
and violence rule the nights, P.S. 81 was in Riverdale, a safe, relatively insulated middle-class community in the northwest corner of
the Bronx.) By midyear, however, on her own time and initiative,
Mrs. Kreindler clipped a small advertisement from the local newspaper announcing that year's offering of astronomy courses at the
Hayden Planetarium. One of them was called Advanced Topics in
Astronomy for Young People, intended for kids in upper junior
high school and the first years of high school. Mrs. Kreindler knew
of my growing interest in the universe based on the alarming pro-

portion of astronomy-related book reports I had submitted. In spite of the age requirement for the courses listed, she presumed they would not be out of my reach and recommended that I explore them. She probably also figured that if my excess social energy was intelligently diverted outside of school, I could grow in ways unfettered by the formal limits of the classroom. Mrs. Kreindler had indeed repackaged and redirected my "social involvement" that she had criticized, leaving me calm and tame in her disciplined classroom. From then onward, the Hayden Planetarium became a much broader and deeper resource for the growth of my life's interests. I had previously known it to be only a place with a beautiful night sky—but the actual universe is much, much bigger.

A student's academic life experience can be constructed from much more than what happens in a classroom. Good teachers know this. The best teachers make sure it happens, and measure their own success as educators not by how many students earned As in their class but by the testimony of whose lives they enriched.

I have kept most of the significant academic documents from my school years, all neatly laid, grade by grade, in a book sensibly called *Neil's School Years* with the *Neil's* neatly calligraphed into the title by the stationer. From grades kindergarten through my senior year of high school, I kept report cards, art projects, exams, and transcripts. I also logged, in designated places, my list of friends, hobbies, and, in the earlier years, what I wanted to be when I grew up. You were expected to choose from a pre-identified list of professions. For boys, the entire list of choices was as follows: soldier, cowboy, fireman, policeman, baseball player, and astronaut. Fair enough. But the for girls, the list was completely different: mother, nurse, schoolteacher, airline hostess, model, and secretary. How a girl could become a mother but a boy could not become a father was a great biological mystery to me at the time, but let's ignore the period sexism. To the list's credit, there was a place where you could write in your own ambition. Beginning in sixth grade, the entry reads "astrophysicist."

My brother and sister each got one of these books too, as a gift from our paternal grandmother, Altima deGrasse Tyson, who, at the time, during the last several years of her life, lived with us in our Skyview apartment. She knew the value of an education and never failed to talk about staying in school and going to college and using these talents to empower you in society. Although she never went to college herself, every one of her five children did. Her middle (maiden) name became my father's middle name, as well as mine. Although deGrasse is French (very loosely traceable to the French admiral who fought in defense of the American colonies during the Revolutionary War, and who was captured and kept under island arrest in the Caribbean), Altima was strongly influenced by the traditions of England and, in particular, the values they place on formal education. She was born and raised on the Caribbean island of Nevis, now a sovereign nation with St. Kitts. Both were once colonies of the British Empire. She carried these values through motherhood, across Ellis Island, and into grandmotherhood.

The Skyview apartments, where I lived during my formative years, contained apartments lettered "A" through "X" in each of twenty-two stories. Comedians and other entertainers like to taunt residents of trailer parks for living in such small quarters. But apart from being bait for tornadoes, living in a trailer park can't be much different from being packed into a New York City apartment building. The building in which I lived had twenty stories, not twenty-two, because the designers of most tall buildings in New York City (Skyview included) succumbed to a bit of superstitious fear and omitted the thirteenth floor. They also left out the fourteenth floor, to preserve the odd-even-odd-even sequence that numbers tend to follow when you count with them, and so that the two elevator banks—one that served the even floors and one that served the odd floors—would not fall out of sequence.

Suburban home dwellers normally assume they have little to envy of urban apartment dwellers. But I can think of at least one exception. During Halloween, the apartment-dweller's bounty from trick-or-treating is without equal in all the suburbs of the land. My friends and I would each fill a large shopping bag of candy in less than forty-five minutes. After an hour and a half we had a full year's supply. And since all roaming occurs indoors, you could trick-or-treat door-to-door in your bedroom slippers. Another advantage of apartment living comes from the height of the building's roof. An elevator ride to the top of the Bronx, with my telescope in tow, gave me an unobstructed view of the horizon at all points on the compass. As far as I knew, Mount Everest had nothing on me. But I had never been west of New Jersey.

Just before I turned twelve my family moved temporarily from the Bronx to Lexington, Massachusetts, an elite suburb of Boston. My father, Cyril deGrasse Tyson, a sociologist and educator by training, had just served six years as a commissioner under Mayor John V. Lindsay during the most turbulent years of the civil rights movement. I am sure a break in the 'burbs did him some good. He had received a one-year appointment as a Fellow at Harvard's Kennedy Institute of Politics, where he was also a research associate in the Program on Technology and Society. We sublet our New York apartment to live in a private home, on a small street called Peacock Farm Road, with a backyard, grass all around, a plum tree nearby, and a small brook out back. Not what you would call urban living.

That excursion to Lexington, my seventh grade in school, happened to be the most successful academic year of my life. I earned straight As and won the school citizenship award (equivalent to that grade's valedictorian). Lots of people get straight As—at least one person per grade, sometimes one per classroom. Having done it once, I am certain that it's one of society's most overrated talents. Adults who had achieved straight As during their years in school typically collect and concentrate among the faculty of academia.

The ascent to a PhD continually sifts the mixture of students so that, at least in physics and astronomy, nearly every research scientist had a straight-A average in high school or college or both. This fact leads to the simple conclusion that practically everyone who is judged "successful" in society, and who is not an academic, was *not* a straight-A student. Go ahead. Ask them. This list includes CEOs of Fortune 500 companies, successful entrepreneurs, inventors, celebrated artists, accomplished musicians and composers, best-selling novelists, award-winning poets, comedians, screenwriters, producers, politicians, Academy Award–winning actors, and of course, professional athletes. So we have created, and willingly support, an educational system that honors the highest grades in class and on exams, but these same perfect grades bear little or no predictive value for those who will actually express the talent that shapes our contemporary culture. Apart from being among the most recognizable people in world history, the boxer Mohammad Ali, who is generally regarded as having a well-below-average IQ, appears in *Bartlett's Quotations*. But the intelligence maven Marilyn vos Savant, with one of highest IQs ever measured, is neither as well known, nor as widely recognized, nor quoted in *Bartlett's*, nor, for that matter, is curing cancer or otherwise researching the secrets of the universe. Last I checked, she had a column in *Parade* magazine that titillates other high-IQ people with word problems and other brainteasers. I hate to be judgmental, especially since Ms. Savant is an easy public target, but that résumé reads a bit thin given what her schoolteachers surely said she would one day become.

My stellar performance in seventh grade was nonetheless a personal achievement because I have never done so well in school, either before or since. I may never know for sure what recipes for discipline made that year unique, but I watched no television and had no playgrounds in view of my bedroom window, just grass and trees. Not to mention that spooky silence at night: no police sirens, no car horns, and no loud voices from people arguing on the street corner. Actually, the nights weren't completely silent. I will not soon forget the annoying cacophonous crickets each evening. I have come to question why these sounds are known as "natural" while sounds made by members of our own species are known as "noise."

I did not grow accustomed to the crickets until I could extract environmental information from their behavior. I deduced for myself the semi-well-known relation between the rate a cricket stridulates and the ambient outdoor temperature: If you count the chirps in fifteen seconds, and then add forty, you get the temperature outside in Fahrenheit degrees. Only when the temperature dropped below forty degrees did the nights fall truly silent.

While in Lexington I also received my first telescope—a birthday gift from my parents. My cosmic interests had already been established, so the 2.4-inch refractor with three eyepieces and a solar projection screen was not one of those "wishful thinking" gifts that parents are known to give their children. The telescope's educational and inspirational value was immediate. And I had a backyard where I could observe the heavens for hours and hours without distractions of any kind. In the daytime I would observe the migration of sunspots across the Sun's differentially rotating surface, tracking their twenty-five-day journey. At night, with the relatively dark skies of suburban New England, the stars and planets were mine. During the snowy Massachusetts winter, I would shovel a path to a circular clearing in our backyard so that I could observe the sky even when the inclement weather left the city at a standstill.

My interest in the universe was in the fast lane, and soon outstripped my telescope's power. All other things being equal, bigger telescopes are better than smaller telescopes. Unlike what you might be told in other sectors of life, when observing the universe, size does matter, which often leads to polite "telescope envy" at gatherings of amateur astronomers. Larger telescopes simply gather more light and see dimmer things. During my formative junior high school years back in New York, I received no weekly or monthly allowance from my parents, although they would not hesitate to buy cheap remaindered books—on math and the universe—that fed my interests. Expensive acquisitions required a job.

I bought my six-inch Newtonian reflecting telescope, brand-named "Criterion Dynascope" from monies I earned by walking resident dogs of the Skyview complex. These weren't ordinary dogs. These were fluffy apartment-dwelling city dogs, not to be confused with the streetwise variety with half an ear bitten off, and that might live in an alley near the Dumpster. I walked large ones, small ones, friendly ones, and mean ones, young ones, old ones, smelly ones, and clean ones. But what they all had in common was disdain for inclement weather and a strong preference for taking the elevator instead of walking up or down the stairs. Going outside was a distraction from their warm and dry apartment life. Most dogs had raincoats, some had hats and booties. I earned fifty cents per dog, per walk, during all my years in junior high school— enough to pay two-thirds the cost of both my telescope and an entry-level Pentax SLR 35mm camera, equipped with specialized adapters for astrophotography. My parents kicked in the rest, fully convinced of my commitment to the subject and the supporting hardware it required.

With its five-foot-long white tube, mounted with counter-weights on a heavy-duty metal pier, my telescope looked like a cross between an artillery cannon and a grenade launcher. Like most telescopes above a certain cost, mine was equipped with an electric clock drive that compensated for Earth's rotation by tracking the motion of stars across the sky. The two-acre roof of my building had no power outlets, but my dentist (a lifelong friend of the family) happened to live on the nineteenth floor. I would faith-fully haul to the roof, along with my telescope, a hundred-foot heavy-duty extension cord that I would unravel across a four-foot brick safety railing, and down into the bedroom window of my den-tist's apartment. This journey to the roof was not easily accom-plished alone. I often indentured my sister, four years my junior, to haul the heavy parts because I did not trust her with the lighter, but much more expensive, optical tube assembly. Thirty years later, my sister still complains about it.

In my midteen innocence, I was simply reaching out to the universe. As for nosy neighbors, my rooftop activities looked to them as though I were a heavily armed burglar, ready to rappel down the side of the apartment building in the dark, with my portable assault weapons strapped to my side. One out of three times I was on the roof, someone would call the police.

Whatever has been said of urban police officers, I have yet to meet one who was not impressed by the sight of the Moon, planets, or stars through a telescope. Saturn alone bailed me out a half dozen times. For all I know, I would have been shot to death on numerous occasions were it not for the majesty of the night sky.

During my junior high and high school years, I attended at least a half dozen courses offered by instructors on staff at the Hayden Planetarium. The subject levels and the expertise of the instructors merged with my stage of learning to create the most formative period of my life.

Among the course instructors, if there were ever a "voice of God" contest, Dr. Fred Hess would win. Dr. Hess is a friendly man, with the body proportions of Santa Claus and a public trustworthiness rivaling that of Walter Cronkite. Both courses I took from Dr. Hess were held inside the Hayden Planetarium sky theater. My favorite class was Stars, Constellations, and Legends. The resonant frequencies of his amplified voice within the planetarium dome somehow conspired to create a fatherly, yet Zeus-like sound that seemed to emanate from the depths of space itself. At a time when my exposure to advanced math and physics classes was growing, Hess's course reveled in the majesty and romance of the night sky, reaffirming for me the simple joys of just looking up.

My current lecture manner and style under the dome of the sky theater, and under the night sky itself, remain traceable to the talents of Dr. Hess. He also happens to be a seasoned eclipse chaser and is among the top few people in the world for total time logged in the Moon's shadow.

For most of the years I attended those Hayden courses, the head of the planetarium was Dr. Mark Chartrand III, an intelligent, committed, and enthusiastic educator who brought a patina of humor to almost everything he taught—not in the form of gratuitous jokes or one-liners, but as a natural flow in his delivery of content. Dr. Chartrand's command of astrophysics mixed with his sense of humor was a combination I had never dreamed possible. If the universe is anything, it should be fun. After meeting a long string of athletic heroes, from track stars to baseball players, I finally met someone who would break the athletic mold and serve as my first intellectual role model. I didn't want to be him, I simply wanted to know the universe and communicate it to others just the way he did. I took two courses from him over the years. My favorite between them, and my favorite of them all was simply titled Astronomy Roundtable, which covered the physics and the mathematics of relativity, black holes, quasars, and the big bang. At age fifteen, I was the youngest in the class by at least fifteen years, and understood, perhaps, only half of the content. But if I were ever going to have the cosmos at my fingertips, I had to begin somewhere.

Back then, the Hayden Planetarium issued certificates for everyone who completed a course. Printed on heavy stock, they resembled graduation diplomas in style and appearance, making them suitable for framing and for remembering. I still have every one I earned, signed by the director. Apparently, the tradition faded over time. But when I became director of the Hayden Planetarium, twenty-five years later, I resumed the tradition, on the premise that this simple gesture may influence the next generation of would-be scientists the way those who came before had influenced me. To efficiently plow through the hundreds of certificates per year that we now issue, my assistant recommended I have my signature made into a rubber stamp. I declined. From my point of view, signing these certificates, one by one, with a fancy fountain pen, is one of the last great privileges of office.

Of all the planets in the sky, my favorite is Saturn. Without question, debate, or argument, Saturn is the most beautiful. My first view through my first telescope was of Saturn. Imagine the thrill of locating a point of light on the sky, then centering it in the crosshairs of a finder scope, and then looking through the telescope's eyepiece to reveal another world—a floating celestial orb surrounded by a ring system three times the width of the planet itself. Several moons are clearly visible through a simple telescope, but at last count dozens are cataloged.

Meanwhile, back in seventh grade, one of the units in wood shop (at the time, still segregated for boys) required everyone to make a lamp. I decided to craft one of my own design even though the class was encouraged to use one of several presketched styles. One of the stock plans was inspired by a water pump, where you press down the pump handle to turn the lamp on and off. Another was a mini wine keg, where the light bulb was where the stopper would go and the switch was the spigot. These designs were clever, and they tested key shop skills, but none resonated within me. My wooden lamp would have a cosmic theme. My wooden lamp would be designed after a planet. My wooden lamp would be Saturn. In my design, the light bulb housing sits atop a lathed, white pine sphere about nine inches in diameter. The power cord passes through a conduit, drilled pole-to-pole through the ball. Two dowels emerge from the equator of the ball to support a broad mahogany ring that tilts on the dowels. With the lamp's chain connected from the base of the bulb housing to the edge of the ring, the lamp turns on and off by tilting Saturn's ring. A wooden pedestal supported the ball from below, with a layer of felt underneath to protect the furniture upon which it rests.

I got an A+, and it remains my primary desk lamp today.

I enjoyed another, albeit obscure, encounter with Saturn while onboard the SS *Canberra* on the way back from viewing the total solar eclipse of June 1973,* off the coast of northwest Africa. With a mobile platform, you are no longer susceptible to inclement or otherwise cloudy weather on the day of the eclipse, provided you carry along a reliable meteorologist. This particular Cunard luxury liner had been converted to a floating scientific laboratory where all manner of astrophysical experiments were conducted during the seven-minutes of blocked sunlight—one of the longest eclipses on record. (A decade later, the ship was converted once again, but this time into a military transport to ferry British troops to the southern hemisphere during the Falkland Islands war.) I had received a small scholarship from the Explorer's Club of New York to take this trip. At age fourteen, I was the youngest unaccompanied person on this fifteen-day trip, but with my telescope in tow, I had all the guardianship I needed. And when people asked my age, I lied and told them that I was sixteen, feeling sure at the time that being two years older would somehow make a difference to the adults on board.

The Explorer's Club is a well-paneled, well-upholstered place on Manhattan's Upper East Side where, in every room, you are sure to have some wall-mounted bodiless mammal staring straight at you. The club attracts, among its members, explorers of every ilk, who go to hang out and share stories of their adventures to the bottom of the ocean, to the tops of mountains, to the depths of jungles, and to the reaches of space. It so happened that Vernon Gray, their director of education, took the Astronomy Roundtable class at the same time I did at the Hayden Planetarium. During the break in one of the classes, after I had asked Dr. Chartrand a flurry of questions about black holes, Mr. Gray walked up to me in a quiet and unassuming manner. He introduced himself, handed me his business card, and invited me to call him when I had the chance. I was

---

*July 1973 coincided with the five hundredth birth of the Polish astronomer Nicolaus Copernicus, the father of the Sun-centered model of the universe.

naive and oblivious. How often does an adult hand a business card to a teenager? It took the diligence of my mother to call the fellow, the very next day, knowing full well the meaning of such a gesture. The Explorer's Club not only awards scholarships to students, but it also maintains a reference database of other scholarship programs for budding scientists to attend expeditions around the world. That brief encounter with Mr. Gray, directly and indirectly, spawned a series of opportunities in my early career—beginning with the eclipse voyage—that stoked my interest in the cosmos, and has left me reflecting how one's trajectory through life can be so influenced by chance encounters with people who wield the power of opportunity.

Two thousand scientists, engineers, and eclipse enthusiasts were onboard the *Canberra*, along with assorted luminaries such as astronauts Neil Armstrong and Scott Carpenter. The supremely prolific Dr. Isaac Asimov was also aboard. He gave a thoroughly entertaining and informative lecture, flavored by his inimitable Brooklyn accent, on the history of eclipses. Although that was the first and only occasion I met him, fifteen years later I reminded Dr. Asimov of the eclipse cruise in a letter, humbly requesting that he write a jacket blurb for my first book, *Merlin's Tour of the Universe*. He agreed, and supplied kind words that any publisher would love. Asimov had read the manuscript and replied within seventy-two hours. Since one of these twenty-four-hour periods was a Sunday, when mail is not delivered, I might have otherwise received the reply within forty-eight hours. Could this busy man have actually read my three-hundred-page manuscript, and typed a reply in this time? I may never know for sure, but he did find, and tell me about, a small error midway through the book.

Four scientists and educators represented the Hayden Planetarium on the SS *Canberrra*, including Fred Hess and Mark Chartrand, who each gave multiple lectures during the cruise. Hess, in fact, served as master of ceremonies of the eclipse itself. There we were. There was I: a thousand miles off the coast of northwest

Africa, and the two leading educators on the boat worked at New York City's Hayden Planetarium. I was a lucky kid.

Apart from the multiple dozens of lectures and presentations, the seven-day journey home included fun intellectual diversions such as an astronomy trivia contest, where my knowledge of Saturn happened to matter greatly. With about fifty contestants, teamed in tables of four or five, an announcer started asking all manner of cosmic queries. An early wave of hard questions swiftly eliminated many tables. One question stumped everyone: "Which day of the year can never have a total solar eclipse?" To that question, I was thinking about dates of the year when I should have been thinking about days of the year. The correct answer was Easter, which is defined to fall on the first Sunday after the first full moon after the first day of spring in the Northern Hemisphere. Easter therefore falls, at most, seven days after a full moon, while total solar eclipses happen only during the moment of a new moon, which falls a full two weeks away from the full moon. In retrospect, any Moon-based holiday that does not specify the astronomical new moon would also qualify as a correct answer such as Good Friday, Passover, the beginning of Ramadan, and Tet, the Chinese New Year.

Another question that stumped, and therefore eliminated, several tables was, "What are the linguistically correct names for objects or aliens from Mars, Venus, and Jupiter?" I knew this one cold. While everybody knew that Mars aliens are called Martians, fewer people knew that Jupiter aliens are Jovians. And only even fewer people knew that aliens from Venus are called Venereals. Astronomers do not commonly use Venereal, in favor of the less contagious-sounding Venutian. Blame the medical community, who snatched the word long before astronomers had any good use for it. I suppose you can't blame the doctors. Venus is the goddess of beauty and love, so she ought to be the goddess of its medical consequences.

At the end of the contest, two tables remained in the running, including mine. The final question was this: "What feature of

Saturn, other than its beautiful ring system, strongly distinguishes it from all other planets in the solar system?" I knew that my Saturn lamp, from seventh-grade wood shop, would float if you tossed it into a bathtub because it's made of wood. Wood is less dense than water. Saturn too would float if you could find a bathtub big enough to place it. Saturn is the only planet whose average density falls less than that of water. I stood up before the assemblage and delivered the winning answer. For that bit of trivia I earned applause from everyone in the room and a free bottle of champagne for my table. Having gazed so long at the stars, I now had my first taste of being one—if only for a brief but effervescent moment.

My second real trip away from my family was during the August that followed my ninth grade in junior high school. Destination: Camp Uraniborg, an outpost in the Mojave Desert of Southern California—directed by two committed science educators, Joseph Patterson and Rick Shaffer—for kids whose parents didn't know what else to do with their precocious progeny for the summer. Uraniborg was the name of the Danish astronomer Tycho Brahe's sixteenth century telescopeless observatory, where, using precision sighting instruments, he made seminal observations of the positions of planets as they moved against the background stars.

Whatever could possess rational adults to lease land in the desert, acquire a dozen high-performance telescopes, assemble a teaching staff of mathematicians, physicists, and astronomers, and invite early teenage astronomy buffs to spend a summer living nocturnally? I don't really know, but Patterson and Shaffer created something special indeed. They created the camp out of a deep love for astronomy and an even deeper love for teaching it to others.

To get there, I joined a half-filled van of others from the East Coast and we drove for fifty-three consecutive hours from New York City to the secluded campsite, thirty miles beyond Barstow,

California. For a month, I lived nocturnally, gained access to a bank of high-performance telescopes, programmed a computer (in 1973, desktop computers were still novel), and took courses in math, relativity, optics, and astrophysics.

I thought I had died and gone to the great sky beyond.

From New York City, on a wishful night, you might spot a hundred stars. That night from the Mojave Desert I saw bezillions. Apparently, my first sky show, six years earlier, was not a hoax after all. With near-zero humidity and dark, cloudless skies, I couldn't help thinking, "It reminds me of the Hayden Planetarium sky," which is an embarrassingly urban thought. That summer I obtained the greatest color photographs I have ever taken of the night sky—before or since. The portfolio includes moons, planets, star systems, galaxies, nebulae, and large swaths of our own Milky Way galaxy. I had captured the soul of the night sky with my Pentax SLR camera on Kodak's high-speed Ektachrome film. All that dog walking back at Skyview paid off again.

For astrophotography at night, high-speed Ektachrome was the most light-sensitive color film commercially available—ideal for the astronomer on the go. During the daytime hours, I used Kodachrome film to capture the desolate desert. Kodachrome happened to inspire a song of the same title by Paul Simon, which received considerable airtime on the radio that summer of 1973. During the cross-country caravan to Camp Uraniborg, a pop-music radio station somewhere in the Midwest bleeped the word "crap" from the opening line of "Kodachrome": "When I think back on all the *crap* I learned in high school . . ."

What country was I in?

Wasn't this supposed to be America, land of liberty and the freedom of speech? At age fourteen I had never noticed that, while our nation is indeed a union, and while we have an interstate highway connecting us all, it's possible for one state to be socially, politically, and philosophically disconnected from the next. Fortunately, the laws of physics apply everywhere on Earth and in the heavens, transcending social mores. These same laws began to serve as one of my intellectual anchors amidst the irrationalities of society.

No less memorable than snuggling with the cosmos that summer was snuggling with the palette of insects and other creatures that claim the desert as their home. The only scorpion I had ever seen was the one of my imagination, traced by the stars of the zodiacal constellation Scorpius. That summer I was shaking them out of my boots each morning—boots I was wearing to protect my ankles from moon-bathing rattlesnakes at night.

Who ever said the desert is a tranquil place? Unlike urban lunatics, Mojave desert coyotes don't need the sight of a celestial orb as an excuse for their unruly behavior—they howled every night—moon or no moon. And there is no doubt about it: bulbous, hairy tarantulas are much uglier, and far more terrifying, than any other creature in the solar system.

After a week's exposure to desert bug fauna, I started longing for the simplicity of urban household cockroaches. They don't sting, bite, suck blood, or inject venom. And they generally stay out of your way.

Camp Uraniborg no longer exists, but the influence upon its participants was indelible. Five(!) of my fellow campers from Uraniborg that summer went on for PhDs in astrophysics, and we all overlapped at one time or another in graduate school. Joe Patterson subsequently received his PhD in astronomy at the University of Texas and is now full professor of astronomy at Columbia University in New York City. Rick Shaffer is a telescope consultant, an author, and a regular contributor to *Astronomy* magazine. Patterson would ultimately serve as a latter-day mentor for me during the transfer of my graduate program from Texas to Columbia.

My experience at Camp Uraniborg remains one of the most enduring and endearing episodes of my life. I was on a path that began at age nine, in the dome of the Hayden Planetarium. My earliest memories of life begin at age four, watching an episode of the *Mickey Mouse Club* on television with my older brother and mother, who was pregnant with my sister. By age fourteen (my age at Camp Uraniborg) my interests in the universe had already occupied half my sentient years.

At summer's end of 1973, my fate was set, having just returned from my one-month session at Camp Uraniborg, where I had obtained striking color photographs of cosmic objects. By then, I had acquired my second telescope and I was a card-carrying member of New York's Amateur Astronomer's Association. Meanwhile, news had broken that the Hungarian astronomer Lubos Kohoutek found a beautiful new comet in the sky. He discovered it much farther out in the solar system than where new comets are typically found, which was a sure indication that its brightness would increase to record levels as it neared the Sun that December.

Comets are basically big balls of dirty ice that can reach a few dozen miles in diameter. They typically orbit in elongated paths around the Sun and contain pristine materials, left over from the formation of the planets nearly five billion years ago. As comets near the Sun, the growing radiant energy changes their surface ice directly from solid to gas, just what happens to dry ice—frozen CO_2—on Earth. The evaporated gases collect around the comet's nucleus, forming an enormous spherical envelope called a coma (Latin for "hair") that can reach millions of miles in diameter. The gases also stream forth into interplanetary space and form a "tail" that gets pushed away from the nucleus by pressure from sunlight as well as from pressure from the persistent stream of particles that emanate from the Sun, called the solar wind. A comet's tail can extend up to a hundred million miles through space but will always point opposite to the direction of the Sun, no matter which direction the comet happens to be moving. That winter, Comet Kohoutek would be easily visible to the unaided eye and was the most anticipated comet in a generation.

During the months that immediately preceded the comet's closest approach to the Sun, I began to see anxiety-ridden people on the street urging others to repent. They claimed the new comet was a sign that the end of the world was near, so now was the time to confess all your sins. How was this profound expression of scientific illiteracy possible? Only two years earlier, *Apollo 14* astronaut Alan B. Shepard Jr. hit golf balls on the Moon. One year earlier, *Apollo 17* geologist Harrison H. Schmitt collected unusual rocks from the lunar surface. And that same year, 1973, NASA launched the interplanetary space probe *Pioneer 11* with a trajectory that would cross the asteroid belt, tour the outer planets Jupiter and Saturn, and then escape the solar system altogether. To my maturing sense of reason, full-grown adults couldn't possibly be so unaware of our technological advances and their attendant scientific truths. Actually, there is no shame in not knowing. The problem arises when irrational thought and attendant behavior fill the vacuum left by ignorance. I knew well that throughout history the arrival of comets, the alignment of planets, and the spectacle of eclipses consistently extracted irrational behavior in people. For example, Comet Halley's arrival in 1066 CE was blamed almost entirely for the Norman conquest of England in the same year. Okay. That was a millennium ago. I just couldn't imagine the prevalence of such behavior in modern times. Maybe that missing thirteenth floor of my apartment building signaled a still-deeper condition in society. Perhaps I could no longer revel in the beauty of the universe without accepting the tandem duty of sharing its laws and operations with those whose superstitions leave them in fear of it.

As word of Comet Kohoutek spread, word of my cosmic interests spread among my extended relatives and family friends. The family network helped in many and varied ways to provide an intellectual buoyancy to my pursuits. One of my mother's many cousins, Francis Crawford, worked in the Queens Public Library where she never failed to acquire and send de-accessioned astronomy and math books my way. A close and lifelong friend of my parents had some expertise in photography and black-and-white-film processing. She served as a first mentor in my early days of astrophotography. Another close friend of the family, who happened to be a professor of education at the City College of New York, recommended me to one of her colleagues who taught at CCNY's Workshop Center for Open Education, a continuing education programs for adults. This instructor, in turn, invited me to give a talk to her Fall classes on whatever topics or aspects of the universe interested me. By then, everybody knew about Comet Kohoutek. With my astrophotographs freshly taken via the lenses and mirrors of desert telescopes, and Kohoutek in the news, I gladly accepted the invitation.

To a class of about fifty people, I devoted most of an hour to describing the subjects of my photo essay—from planets to stars to the Milky Way galaxy, ending with a special discussion of the lore and science of comets and what the winter sky will look like with Comet Kohoutek as a visitor. In this first lecture of my life, I wasn't the slightest bit nervous, even though the room was filled with people who were two, three, and four times my age. For me, talking about the universe was like breathing. I suppose it was no different than another kid talking about his treasured baseball card collection, or a film buff recalling scenes from a favorite movie. I could not have been more comfortable sharing what I knew.

Three days later I received a check in the mail for $50, with a request to return and give two more lectures. Apparently, this was their standard compensation for visiting speakers. At the time, minimum wage was $1.60 per hour and I was two months past my fifteenth birthday. Given these two facts, $50 looked like a semi-infinite amount of money for an hour's work. In dog units, that's one hundred walks.

After landing back on Earth, I felt like an information prostitute. I had never before been paid to speak—paid to share information that just happened to be lying around in my head. Could the act of helping or enlightening others be classified as remunerable labor? Imagine volunteering to help a little old lady across the street, and when you get to the other side she pays you for your efforts. You didn't do it for the money. You did it because it was the right thing to do and because it felt good to help another person. How dare she even make an offer. Of course, I didn't return the check. My fleeting feelings of financial morality were replaced with the lesson that knowledge and intelligence were no less a commodity than sweat and blood.

That December, when Kohoutek was at its brightest, the comet was barely noticeable to the unaided eye. Most people had to be told where to look. Kohoutek was a dud. The scientific community would learn for the first time that comets with orbits lasting hundreds of thousands to millions of years (like that of Kohoutek) tend to evaporate inefficiently and, as a result, produce tiny comas and tails. Fortunately, I was not asked to return the $50.

The following summer, I went on a two-week trip with Educational Expeditions International to the outskirts of Kilmartin, a teeny town near the west coast of Scotland surrounded by fertile green farms. The education office of the Explorer's Club, through the efforts of Vernon Gray, had alerted me to yet another opportunity. In Scotland, I joined a team of scientists and surveyors to excavate and map the astronomical alignments of uncharted prehistoric megaliths, not unlike those of Stonehenge, although much less celebrated. The monies came from a grant by the U.S. Department of Education, Office of the Gifted and Talented. My application survived the several levels of competition, from local, through city, and then state, until the authorities beknighted me "gifted and talented," a designation I resented then and now. Was I born with my expertise? Did somebody hand me my expertise? Who gave me my talents as a "gift"? Hadn't I worked hard and long to achieve my expertise? Hadn't I received support from loved ones who cared deeply about my ambitions? The title "gifted and talented" specifically disavows people who fall below the arbitrary threshold for such measures. A more appropriate, though less catchy title might be the Department of Education's "Office of the Students Who Work Hard." This title would, instead, challenge the nonwinners to work harder and do better next time, rather than give up for not having been deemed gifted.

After a fortnight of surveying and analyzing the astronomical abilities of prehistoric residents of the British Isles, I headed back home. As I passed through London's Heathrow Airport, I spotted a newspaper headline announcing that President Nixon had resigned over flak from the Watergate Hotel break-ins. I had been abruptly cast back into the "real world," although I preferred to think that it's the universe and not Watergate that holds this distinction.

In my senior year at the Bronx High School of Science, I was elected captain of the wrestling team. But I was also editor in chief of the 1976 *Physical Science Journal*, which ranks, along with the editors in chief of the annual *Math Bulletin* and *Biology Journal* as the most prestigious title one can hold in the school (rivaling, perhaps, the mystique held by the quarterback in most other schools). I was proud of that volume. It featured my field report from Scotland, as well as a dozen other original research articles from my classmates. Titles included "Nonlinear Stensor Analysis," "Using Hafnium-182 to Treat Malignant Melanoma of the Conjunctivae," and "The Determination of the Technological Feasibility of the Photon Rocket." The cover page portrayed the plaque that had been affixed to the side of both *Pioneer 10* and *11*, the first space probes to acquire enough energy to leave the solar system and enter interstellar space. The plaque displays a line drawing of a nude man and woman along with a variety of cryptic symbols and signs intended to give intelligent aliens our location in space along with a hint of our scientific station in life. The end pages of the journal contained an assortment of physics brainteasers while all pages used the sequence of element symbols from the periodic table to designate page numbers. You can't get nerdier than that.

At sixty-four pages, that edition of the *Physical Science Journal* was, at the time, the largest ever produced at the school in any subject.

In the fall of my senior year, I applied to five colleges, including Harvard, MIT, and Cornell, which were my top three choices. As a courtesy to applicants, the eight Ivy League schools, plus MIT, notify you by midwinter about whether your application is unlikely, possible, or likely to gain you admission to the school. The first of these to arrive was from MIT, on a day when I just happened to be retrieving the mail. While standing in front of my open mailbox, I held the envelope up to a beam of sunlight, which was piercing the mailroom's window (as though if I had opened the envelope quickly, the result would magically change for the worse). That's when I glimpsed the word LIKELY, circled boldly in red, and that's when I knew the next chapter in my life was set. That moment represented the greatest emotional swing I have ever experienced— from a state of high anxiety in one instant, to a prostrate and teary-eyed state the next.

For most of my high school years, I subscribed to *Scientific American*. The "About the Authors" section was my favorite because it contained all sorts of personal information about the contributing scientists, such as where they went to school and what their side interests are. One prominent astrophysicist, the late professor David Schramm, had also been national Greco-Roman wrestling champion. So there was at least one other wrestling scientist out there. When choosing a college to attend, I devised a decision matrix that tallied the number of physics and astronomy articles in *Scientific American* written by scientists who as undergraduates attended the schools that admitted me. I also tallied where these same authors earned their master's degrees, their PhDs, and on whose faculty they currently served. Harvard won in every category, although Cornell University represented a strong draw for me because of Professor Carl Sagan's presence on their faculty.

I first met Carl (as he preferred to be called) during a visit to Cornell for the required college interview. My letter of application had been dripping with an interest in the universe. The admission office, unbeknownst to me, had forwarded my application to Carl Sagan's attention. Within weeks I received a personal letter inviting me to visit him in Ithaca, New York, the secluded home of Cornell University. Was this, I asked myself, the same Carl Sagan who I had seen on NBC's *The Tonight Show*, with Johnny Carson? Was this the same Carl Sagan who had written all those books on the universe? Indeed it was. I visited Cornell on a snowy afternoon in February. (I later learned that many winter afternoons in Ithaca are snowy.) Carl was warm, compassionate, and demonstrated what appeared to be a genuine interest in my life's path. At the end of the day, he drove me back to the Ithaca bus station and jotted down his home phone number—just in case the bus could not navigate through the snow and I needed a place to stay.

I never told him this before he died, but at every stage of my scientific career that followed, I have modeled my encounters with students on my first encounter with Carl.

I did not ultimately attend Cornell University, because the data from my analysis of *Scientific American* authors was too compelling to forego. So I was off to Harvard, but not before my character and my manhood would be tested.

During my senior year of high school, just as spring began, New York City received one of those winter's-last-stand snowfalls that dumped four or five inches across the metropolitan area. Spring snows tend to fall at temperatures just below freezing, sometimes at or above freezing, which makes for wet snow—the kind that sticks to the thinnest of a tree's bare branches and the kind that makes the best snowballs. That day the sun was bright, made brighter by reflecting off the freshly fallen snow, and the temperature was rising toward forty degrees. School wasn't canceled but it might well have been. The snow would not last twenty-four hours before succumbing to the spring sun, and most seniors, myself included, took very long recesses (i.e., we cut classes) that day to engage in a schoolwide snowball fight. Off to the side, leaning against their motorcycles, stood several students who never went outside without donning their black leather jackets, each with its own configuration of metal studs, zippers, and chains threading its epaulets. They were the "greasers" and represented the toughest faction of the school. I cannot judge how tough they would be at an average city high school, but at the Bronx High School of Science, they were mean and scary. During the snowball fight, the greasers that day were noncombatants. But one of my high arching snowballs, one that sailed probably fifty yards, happened to veer off course, like a sliced golf shot, and hit squarely on the chest of the leader. As the snowball exploded into his chest, it scattered snow into his face with his girlfriend in full view, tucked under his arm.

A most unfortunate incident.

Anybody with less testosterone would just have laughed it off. But the head greaser immediately yelled all manner of expletives and racial epithets across the yard to me, with fist clenched and waving in the air. While sound does not travel well across fields of snow, I heard him clearly say that I had better not walk past him and return to school that afternoon, under threat of bodily harm.

An hour later, at the end of the snowball fight, most people were cold, wet, and tired—the perfect occasion to return to whatever class we belonged in. As I slowly, but confidently, approached the school, in defiance of his previous commands, the angry greaser walked toward the main door at a pace that would perfectly intersect my arrival. When I drifted, instead, to the slightly farther side door, he quickened his pace. When we came within about ten feet of each other, he drew his "007" pocketknife, which had a strong wooden handle, a six-inch locking blade, and could be opened as swiftly as a spring-loaded stiletto. We stopped within two feet of each other just when he positioned himself between me and the school door. We had never before stood so close. I think he was slightly surprised that I stood two inches taller, including the one-inch heels of his motorcycle boots. He may also have been surprised that while I mildly tried to avoid him, upon reentering the school, I neither flinched nor showed signs of fear when we stood face-to-face.

He wielded his knife in the bright sunlight, with the Sun's image reflecting back and forth across my face. I thought this sort of thing happened only in the movies, or in highly choreographed urban fight scenes, like what you would expect to find in the musical *West Side Story*. Holding his knife within two blade lengths of my face, he uttered softly, "You hit my leather jacket with a snowball." I next compressed what would normally be fifteen minutes of logical reasoning into probably two or three seconds of silence. My timed mental flowchart went something like this:

T= 0–1 seconds
    He wants to engage me in a fight.
    He's brandishing a large, sharp knife and I am not.
    I am captain of the wrestling team and I am bigger than he is.
    I am probably also quicker, and I have some training in martial arts.
    I could probably disarm him and pin him to the ground.
    But then he would lose the fight and be angrier still.

T = 1–2 seconds
    He and his friends would surely seek revenge before year's end.
    I would live my final months of high school in fear and terror.
    But suppose I fail to disarm him?
    However small that chance is, I may be cut badly and possibly die.
    I have much more to lose in this fight than he has to gain.

Harvard, the college of my choice, has already accepted me.
My entire life of astrophysics lies ahead.

T = 2–3 seconds
My identity flows from neither my ego nor my virility.
After all, it was *my* snowball that hit him.
Mahatma Ghandi and Martin Luther King Jr. praised nonviolence.
If I humbly refuse to engage, he has no force to combat.
With no force to combat, he may just disarm himself.

Like the one-line output of a long computer program, my
mouth promptly uttered the following sentence, with sincerity: "I
am sorry I hit you with a snowball. It was unintentional and will
not happen again."

We stared at each other in silence for several seconds more
before he folded up his knife, returned it to his pocket, and walked
back to his motorcycle. I quietly passed through the school's side
door and returned to class, having retained my health, my dignity,
and my future.

The Middle Years

There's much written about the value of a broad formal education to
one's enlightenment. But when all is learned, a double standard
remains. At cocktail parties, if a conversation touches upon late-
nineteenth-century literature, or baroque music, or renaissance art,
then the participants get tagged as "erudite." But if the conversation
were about a hadron supercollider or about the hydrodynamics of
dams, or about the emergent market for hydrogen fuel cells, the participants get tagged as "geeks." Nobody ever passes judgment on
those who admit, "I was never any good at math." People just accept
such statements, and even chuckle among themselves for having
said it. But just look at people's reactions if someone were to con-

fess, "I was never any good at nouns and verbs." Or "books have just too many words in them." I am occasionally (though playfully) chastised for not knowing some character or another from a Shakespeare play, or from the president's cabinet. Yet, in spite of these double standards, I have come to realize that whatever I know that isn't science, I know far more of it than the science known by nonscientists.

At the Bronx High School of Science, the environment bred logical and analytical minds, so I had not imagined that a liberal arts component to my formal education would amount to anything more than an academic curiosity. I have since benefited greatly from my nonscience training, especially in college, where more than half my coursework fell outside my field of concentration. It all started my freshman year, when I took a course titled Humanities 15. In this full-year survey of art and design, I didn't just learn about art and design, I actually did it.

And I have never been the same since.

The course was taught in one of the studios of Harvard's Carpenter Center for Visual Arts, a hypermodern building designed by Le Corbusier—his only building in the Western Hemisphere. The professor was Louis Bakanowski, one of the founding partners of the Cambridge Seven Associates, a well-known architecture and design firm in Cambridge, Massachusetts. All the ingredients were in place, and I was in the belly of the beast.

Charcoal drawing led off the syllabus at the beginning of September. The instructor first played recorded music of various genres and asked us to draw the music's energy. Excuse me. You want me to do what? Energy is mc^2 or $1/2\ mv^2$ or mgh or $G\ m_A m_B\ /\ r$. Energy is *not* something you draw while hearing music. One of the hallmarks of science is its precision of language and concepts. What else could I think of the assignment but that it was a waste of my time, my tuition, and of course, my energy.

By mid-September we drew nudes, which interested me much more than drawing music—although I think the human body is overrated as a thing of beauty, especially when compared with truly heavenly bodies. By the end of September, we were drawing miscellaneous objects piled in the front of the studio—rocking chairs, drapes, balls, foot lockers—stuff you might find in somebody's attic.

Come October, we were drawing a heap of gnarly pumpkins. By month's end, I must have drawn a thousand of them. I was fluid. I was focused. I was getting better and better at it. In the end, I dreamed pumpkins in my sleep. All was well in Humanities 15 until the instructor commanded, "No longer draw the pumpkins. Draw the space between the pumpkins." At that moment, as best as I could figure, my rational mind snapped.

After a month of pumpkin worship, these things all of a sudden became boundaries to the absence of pumpkin, in which I was to endow the same level of meaning and existence that I had previously granted the pumpkins themselves. To the instructor, I must have looked like a dog that just heard a high-pitched sound as its head tilts with puzzlement.

I eventually became pretty good at drawing the nothing between the something, and I would never look upon the human universe the same way. My private world transformed overnight, from one occupied by all things chemical and physical to one occupied by the juxtaposition of shapes and forms. I broke free of the logic box that I did not know had contained me. From then onward, I welcomed all manner of verbal abstractions and creative use of vocabulary into my life. From it, I continue to derive insights into art, literature, music, and the human condition. I encourage the liberal artists of the world to take a conjugate excursion through the land of logic. For one to thrive in the real world probably requires mastery of both.

My entering class at Harvard contained your usual list of children of luminaries, with Caroline Kennedy and her late cousin Michael Kennedy topping the list. The Harvard student body remains steeped in social traditions that derive from the power of wealth and political influence. The father of a good friend in my dormitory was the governor of Puerto Rico. The father of the wrestling team captain during my sophomore year was the Speaker of the House of Representatives. And one guy came from a town in Maine whose surname was the same as his.

But I was indifferent to it all.

Many, perhaps most, people who attend Harvard do so to share in these Ivy League legacies. I was simply there as my next step in becoming a scientist, and not much else mattered. Neither the football games, nor the student protests (trying to get Harvard to divest its portfolio of holdings in companies that do business with apartheid South Africa), nor the social rituals. And yes, I confess to actually saying "Not Yo-Yo Ma again" the fourth or fifth time he gave a cello recital in one of my dorm's common rooms during his senior year.

One of the few lasting elements of Harvard's traditions on my life appears as a double inscription over one of the ivy-coated arches from Massachusetts Avenue into Harvard Yard, where I lived as a freshman. As you pass under the arch from the street, the inscription reads, "Enter to Grow in Wisdom." Three years later, I would discover its companion inscription, barely visible as you leave the yard on the other side of the arch, "Exit to Serve Better Thy Country and Thy Kind."

My list of life's most influential mentors continued to grow. In graduate school at the University of Texas at Austin, where I earned my master's degree in astronomy, I was a teaching assistant for nearly all the semesters I was enrolled. This arrangement served three needs. I became eligible for in-state tuition, the department got much-needed help to manage and run its battery of labs and introductory astronomy courses (UT has one of the largest astronomy programs in the nation), and I acquired teaching tips from the professors.

Not many professors in this world actually care whether the lecture they deliver is the same as the lecture absorbed by the attending students. For the two to be the same requires a certain level of sensitivity to how student minds can misinterpret what you tell them. My own teaching methods were honed and refined after working for Professor Frank N. Bash. He remains the only professor I have ever seen who teaches to the mind of the student and not to the syllabus or the chalkboard. He eschewed multiple-choice exams, which, of course, significantly increased my burden as a grader; instead, he stressed verbal logic of the kind inherent in well-posed scientific problems. By the end of every introductory astronomy course he taught, the students knew how to think about the physical world around them. Other than Professor Bash's course, I know of no other in all of academia where students who got Cs would still claim that the course was the best they have ever taken. I am a better teacher, a better professor, and a better educator for my time spent as a TA under Professor Bash. For a while, some years later, he and I both happened to serve on the board of directors for the American Astronomical Society, the professional organization of the nation's astrophysicists.

On the graduate research side of the academic fence, one of several advisors on my master's thesis, at the University of Texas, was the late Professor Gerard deVaucouleurs, one of the last of the old-world scientists. He knew Russian, German, French, and a little bit of Latin, all of which gave him access to the historical, untranslated scientific literature. He had an encyclopedic knowledge of all published research relevant to his own. And he was the most meticulous scientist I have ever seen. He logged four or five comments, suggestions, and criticisms per page on the 130-page draft of my master's thesis. Other committee members logged four or five notes per chapter. More than anyone else, deVaucouleurs had instilled within me an uncommon sense of patience, precision, and scientific fortitude. No idea is too big to tackle. And no detail is too small to spend days or weeks investigating. Like the rare and cherished Stradivarius violin, I fear that his kind will never again be made.

I also attended a course taught by Professor John Archibald Wheeler, who is generally credited with inventing the term "black hole," where a blob of matter has collapsed upon itself, closing off the surrounding fabric of space and time from the rest of the universe. A former student of Albert Einstein's, Professor Wheeler, in all his brilliance, remained humble in the presence of Nature as if the laws were a ladder that we all must climb. We stand together at its base, taking measure of the ladder's height and how hard it is to climb. Wheeler was humble about what he knew and honest about what he did not know, leaving him quick to admit an error. He always carried a supply of pennies in his pocket when he taught his graduate physics classes. If you caught him making a mistake on the chalkboard, he would stop the class and publicly hand you one of these pennies. We all should live by these deep, yet simple, philosophies, but they are especially rare among leading scientists.

I also happened to meet my future wife, a physics graduate student, in Professor Wheeler's class on general relativity.

My take-home stipend as a teaching assistant came to about $6,000 per year, and I paid $400 per month for a small one-bedroom apartment, which left about $3 or $4 per day on which to live. A budget such as this strongly limits one's food choices. Assorted combinations of pasta, rice, beans, pork neck bones, eggs, canned tuna, bread, and cheese can get you far. But I remained athletically active after college and my demand for calories outstripped my capacity to feed myself on $3 per day. For the first time in my life, I was unintentionally losing weight.

I needed another source of income.

I don't know why, but the first thought to come to mind was to become a male stripper at a nearby women-only nightclub. The joint was only a mile away, located halfway between where I lived and downtown Austin. I was relatively flexible for my size (six feet two, 190 pounds) having been a performing member of two dance companies while in college, and I was in pretty good shape, having wrestled varsity in NCAA Division I. I could do easy flexible poses such as sit in a full lotus position, and, while standing with straightened legs, bend over and palm the ground with both hands. Commensurate with my training, however, I could also execute harder moves like a side-to-side leg split. I could put my foot behind my head while seated. I could grab my instep and raise either of my legs straight up over my head while standing; and I could curl backward from a standing position until the back of my head touched my heels. From what I had seen and heard about these clubs, men who can bend do better than men who cannot.

I figured I could work at the club for just one or two nights per week to supplement my income. Of all the legal ways to earn fast cash, such as giving blood or making donations to the sperm bank, the nightclub concept intrigued me most. One evening, I decided to stop by and observe in detail the dance sets and the performance rituals of the club's stage. In addition to one's own pelvic gyrations, several routines choreographed for the whole group had become club mainstays. One routine, the finale, required that all dancers wear only a jockstrap. But these were not ordinary jock straps. These were specially designed with an asbestos lining that had been soaked with lighter fluid. Upon igniting them—yes, upon igniting them—the dancers sprung onto the stage shaking their buns and their flaming privy housings to the 1958 Jerry Lee Lewis hit "Great Balls of Fire."

I thought calmly to myself, "Maybe I should be a math tutor."

To set one's genitals on fire seemed more like the absence of a creative solution to money problems rather than the need to dance. And so it came to pass that I tutored undergraduates in math and physics through a campus organization that ensured a continuous supply of students in need of help, and a continuous supply of money.

I am embarrassed for this (non)episode. I should have known well in advance of my strip club excursion that tutoring was the way to go. While in college I had made weekly trips to a maximum-security prison as a volunteer math tutor for prisoners seeking their high school Graduate Equivalency Diploma. The prison was Walpole State Penitentiary, near Walpole, Massachusetts, which had a death row—Cell Block Ten—where an electric chair was housed. My "student" prisoner was doing time for breaking and entering, and he had been in prison before. His name was Carlos, and he was quick to show me the scars from several bullet holes the local police put in his body before he was apprehended on his last convenience store robbery. He had done okay on the reading sections of the GED exam, but his math skills were poor. He needed help with fractions and their arithmetic manipulations. These weekly visits forever changed my view of prisons and prisoners.

After a walk through a magnetometer sensitive enough to detect tooth fillings, you next passed through a set of double steel doors that bookend the inner, reinforced concrete wall. The passage through the wall remains in full view by a downward-looking guard tower that sees through a wide hole in the passageway's ceiling. The prison guards all wore miniature jeweled handcuffs as tie clips. Prisoners convicted of sex crimes were not particularly safe on the main floor of the prison. Within the list of all felonies lives a hierarchy of crimes for which you would earn either the respect or the ire of other inmates. In this subsociety of castoffs there was no crime more noble than killing your wife or girlfriend for cheating on you. Bank and store robberies were also up there in the rankings, especially if you had to fight security guards or other armed personnel. Deep at the bottom of the list one finds child molesting, rape, and other crimes where the victim is helpless and there was no provocation. Without formal protection by prison guards, the secondary trial and sentence passes swiftly. For example, in 2002, John Geoghan, a Roman Catholic priest, was convicted and sent to prison for a single instance of child rape, but only after having been accused of molesting more than 130 people. When I learned of this doubly offensive verdict, I counted the days that he would stay alive. Although held in a special protective unit of the prison, seven months later, Geoghan was bound, gagged, and strangled by an inmate who was already serving a life sentence.

In this parallel universe, with all its own rules, Carlos's crimes sat relatively high in the prison pecking order. I got to know him fairly well. I was nineteen at the time, and he was in his upper twenties, although he looked much younger. He had a fair complexion and a young-looking face with eager eyes. At five feet seven, he was small for a street thug, but he talked tough and wielded a strong urban accent. I suspected that "Carlos" was not his given name because I found no trace of a Hispanic accent, manner, or culture within him. I have no proof, but I bet he chose the name simply to add to his tough street image. Between the math lessons and during casual time reserved for idle conversation, I learned that he played the guitar and loved pop-jazz music. His favorite album of all time, which also happens to be one of my favorites, is Marvin Gaye's classic *What's Going On*. The album, released in 1971, was one of the first concept albums of the genre, where both sides of the LP contained a continuous sequence of songs that called for peace, social justice, love, and reconciliation.

Nothing was gained by formally tutoring prisoners who were on death row, or who were serving life sentences with no chances of parole. They mostly wanted a chess companion or someone to talk to—for many, their families had long ago abandoned them. The several lifers that I had met all managed some kind of benign hobby. One grew plants. One cared for goldfish. Another was writing his life story. The surreal paradox of a murderer who cares for goldfish in his prison cell moved me. Contrary to my naive suspicions, a prisoner can indeed sustain a modicum of civility and quality of life even if he took someone else's life and even if he has no prospect for rejoining society. I valued this fact and vowed to pay whatever extra taxes society levies to sustain those with life sentences rather than execute them.

We may never understand the mind of a societal misfit. But for a brief time, I sat a little bit closer and was heartened to learn that beneath all the crime and the punishment, a small slice of ambition and humanity remains in sight. Sometimes I need to remind myself that for all my scientific understanding of the stars, people will forever see them as poetic hooks upon which to place their dreams.

The Later Years

The decade of the 1990s happened to enjoy several comets of remarkable note, including Comet Shoemaker-Levy 9, arguably the most famous comet in history. Its fame derives not from how bright it was or from how many inspired poets wrote about it, but from its brief and spectacular encounter with the planet Jupiter. At over three hundred times the mass of Earth, and at over ten times our diameter, Jupiter retains an unmatched ability among the planets in the solar system to attract comets. During the week of celebrations for the twenty-fifth anniversary of the *Apollo 11* Moon landing, Comet Shoemaker-Levy 9, having been crumbled into two dozen pieces during a previous close encounter with Jupiter, slammed, one chunk after another, into the Jovian atmosphere. Because Jupiter rotates quickly (once every ten hours), each piece of the comet plunged into a different location of the planet as the atmosphere rotated by. The gaseous scars were seen easily from Earth with ordinary backyard telescopes.

In this game of interplanetary billiards, by far the most dangerous impactor is the long-period comet, which are those with periods greater than two hundred years. Representing about one-fourth of Earth's total risk from all possible impactors, long-period comets fall toward the inner solar system from great distances and achieve speeds in excess of 100,000 miles per hour by the time they reach Earth. A trip from New York to Los Angeles at that speed would take all of ninety seconds. Long-period comets thus achieve much higher impact energy for their size than your run-of-the-mill asteroid. More important, they remain too dim over most of their orbit to be reliably tracked. By the time we discover a long-period comet headed our way, mere months would remain to fund, design, build, and launch an interceptor that would save Earth and all its ground-dwelling inhabitants.

For example, in 1996 the Japanese amateur astronomer Yoshii Hyakutake discovered a comet while searching with a tripod-mounted, jumbo pair of binoculars. More than anywhere else on the sky, comet hunters of the world search along a band in the heavens that traces the plane of the solar system. All things considered, the plane is where the action is—planets and asteroids and many comets orbit the Sun close to this plane. But Comet Hyakutake came upon us like a tomahawk from nearly ninety degrees above and outside of the plane—while nobody was looking. Comets gain considerable speed as they near the Sun, but when viewed from afar, looming in Earth's night sky, this rate instead looks rather stately, just as a distant, fast-moving airplane appears to move only slowly overhead, even when cruising along at 600 knots.

Bright comets typically enjoy at least a year of hype before they become visible to the naked eye. After Hyakutake was discovered, only three months would pass before the comet reached its closest approach to Earth—about ten million miles—one of the closest comets on record. This hairbreadth distance rendered the comet large and visible in the night sky. The comet could even be seen (with the unaided eye) from the middle of light-polluted Times Square in New York City. After the disappointments of Kohoutek of the 1970s and Halley of the 1980s, we were finally rewarded with a "once-in-a-lifetime" event.

A year after the 1994 impact of Shoemaker-Levy 9 into Jupiter's atmosphere, another comet of high anticipation came around the Sun. Comet Hale-Bopp had been independently discovered by the professional astronomer Alan Hale and the amateur astronomer Thomas Bopp two years earlier, when it was farther away from Earth than any previous comet had been discovered. We knew Hale-Bopp had to be a big comet for it to be seen so far out in the solar system, but would it be bright? Checking in at over twenty miles in diameter, its nucleus was the largest ever seen. As it neared Earth and the Sun, Hale-Bopp got brighter and brighter until it too became a "once-in-a-lifetime" comet. Hale-Bopp broke all records for remaining brighter than the detection limit of the unaided eye for longer than any other comet on record.

During the several months that Hale-Bopp was at its prettiest, I happened to be taking a cross-country flight in the early evening from New York City to Los Angeles. While cruising at thirty-seven thousand feet above sea level, I looked outside my coach-class window and saw Hale-Bopp, bright and beautiful, quietly suspended in the dusk sky. This particular daily flight chases the setting Sun in the west. Even at a speed of six hundred miles per hour, you will lose this race—the "ground speed" rotation of Earth at the latitude of Los Angeles is about nine hundred miles per hour. The plane nonetheless travels fast enough to greatly prolong the majesty of twilight and anything suspended within it.

The sight of Hale-Bopp so moved me that I wanted to share my excitement with all two hundred people on board the Boeing 767. Since pilots love to interrupt tranquil flights with miscellaneous announcements, I knew my efforts would be best served if channeled through them. I carefully jotted down a dozen factoids about the comet. Aside from the basic facts of how and when Hale-Bopp was discovered, I took the risk of including one or two apocalyptic facts. For example, the comet is somewhat larger than the asteroid that took out the dinosaurs 65 million years ago. If Hale-Bopp were ever to hit Earth, humans would become extinct, too. As it penetrated Earth's atmosphere, Hale-Bopp would first create a blast wave that incinerates over one hundred thousand square miles of vegetation surrounding ground zero. Next it would hit Earth's crust, squashing like a bug anything beneath it. Next it would leave a four-hundred-mile diameter crater. The time from first contact with

Earth's upper atmosphere until the full excavation of a crater takes about ten seconds. The excavation of the crater would thrust a trillion tons of dust into the stratosphere, plunging Earth into darkness, knocking out the base of the food chain, and rendering over 90 percent of the world's species extinct.

To this page of jotted information I attached my two business cards, one as director of the Hayden Planetarium and the other as research astrophysicist at Princeton University. I pressed the call button, handed the neatly folded lesson plan to the flight attendant, and said, "Please deliver this message to the pilot." I can't explain why, but that was awfully fun to say. It probably ranks with the act of slipping a note of no consequence to a bank teller.

Sure enough, about five minutes later, the pilot made a planewide announcement that an astrophysicist onboard had supplied him with information about Comet Hale-Bopp. Part of the crew's job is to ensure the safety of its passengers, which, I think, normally includes their mental safety as well. I was therefore delightfully surprised when the pilot read aloud every line of information from my folded note, including the part about a comet impact having the capacity to cause the extinction of our species. Apparently, everybody was so intrigued by the information that the flight attendant came back down the aisle, invited me to sit up in first class, and served me a mini bottle of champagne followed by an ice cream sundae. Not since I was fourteen had I been unexpectedly paid and treated to champagne for willingly sharing my knowledge of the universe.

Most of my pedagogical excursions in my life have been with students (junior high through college) and the general public. Only rarely do I get the chance to talk to teachers, although I love nothing more. Apart from generally being an enthusiastic and friendly lot, they shape the conduit of our nation's brain trust.

Along the way, they work in the trenches while the rest of us sit at home with a TV remote in our palm and bark out complaints about the state of the educational system. The nation's teachers are collectively underappreciated, underrespected, and underpaid, but they are not all created equal.

When I was asked to give a keynote speech in Washington, DC, to the 1998 winners of the Presidential Award for Excellence in Mathematics and Science Teaching, I gladly accepted the invitation. I owed these star elementary schoolteachers all I could possibly give. That evening, I brought with me *Neil's School Years* because I had a thing or two to get off my chest, and I needed documentation.

When I was in kindergarten, during one of the weekly art periods where whatever you drew was destined for a magnetic mount on your home's refrigerator door, I drew a nighttime scene, engaging the black crayon to depict the night sky. (For years I never knew what to draw with that pink-colored crayon called "flesh.") Upon seeing my creation, the teacher politely insisted that I should have drawn the night sky dark blue. I was not swayed, and left my teacher with the burden of proof. Two days later, she conceded the nighttime sky was indeed black after investigating the problem further. This was the first time I can remember when I was right and the teacher was wrong, which is far more severe than the teacher simply not knowing. I had not expected such an incident to happen until at least junior high school.

My fifth-grade geography exam contained a two-part question, "What is the smallest continent?" My correct answer was *Australia*. The follow-up question asked, "In what hemisphere is the smallest continent?" I wrote *Southern* and it was marked wrong. The "correct" answer was *Eastern*. Now there's a teacher who would have never been eligible for the Presidential Award for Excellence.

Everyone in that room, and every student who ever was, can tell stories of teachers in their lives who possessed a singular talent for inspiring our curiosity in ways unmatched by all other forces in society. To me, the least impressive teachers are the ones who bring forth the smartest children from their class—you know the ones, they get straight As and win all the science fair contests—and claim that their good teaching had something to do with it, which cannot be logically true. Students who get straight As can do so without the help of the teacher, in spite of what teachers tell themselves. Indeed, the difference between one teacher's talent for teaching and another's is irrelevant to a straight-A student. That's what it means to get straight As—your performance does not correlate with who's doing the teaching. Instead, the most impressive teachers to me are those who inspire a failing student to become a passing student, or who inspire a C student to become a B student, and, in all cases, spark an interest to learn more than the school curriculum. Without this goal, students become academic automatons, where the joy of learning is sapped, and where grades matter more than insight and ambition.

No matter who you are, support for your interests is always good, but may it not be forthcoming from anyone other than friends and family. In any case, support is best received during times when you think less of yourself than your talents deserve. To be praised for meager or noncompetitive talent, just because you are loved, does you a disservice in any meritocracy, such as the society in which we live. Whenever that happens, you risk leading a deluded life, where the correspondence between what you deserve for your efforts and what you think you deserve is lost. The value of this "reality check" cannot be overemphasized. Neither Cyril deGrasse Tyson, my father, nor Sunchita Feliciano Tyson, my mother, received formal mathematical, scientific, or technical training in their lives. My father is a sociologist and my mother, who returned to school after she raised her children, earned a master's degree in gerontology. Both of them served as a political and social reality check for me, yet they could not directly provide a reality check on my science.

Knowing this, they did the next best thing. They nurtured my scientific growth.

I must have had the first ever "soccer mom," except the activity wasn't after-school soccer, it was after-school astronomy. With my telescope, camera, and other observing accessories, I would drag both of my parents (separately and together) in and out of cars, up and down stairs, in and out of fields, and to and from the library, all in the support of my astrohabit. I will not soon forget when I was building my wooden Saturn lamp in seventh grade. My mother and I drove to at least six different hardware stores one afternoon just to acquire the necessary, but unusual, electrical conduit that threads pole to pole through the wooden orb.

Furthermore, most weekends we would visit one of the city's many museums, and my parents were always on the lookout for affordable math and science books. They might have said no. They said no to plenty of other kinds of requests. Had they said no to that which promoted my intellectual growth, my own ability to evaluate what I had previously learned would be compromised, and I would lose the capacity to exercise a reality check on myself. This capacity is what empowered me to know when my talents exceeded the levels to which they were judged to be by others, and, conversely, it allowed me to judge the times I was the recipient of gratuitous praise.

My parents never told me where to go or what to learn, which ensured that my life's interests were as pure as space itself. To this day, my parents remain two of the most warm and caring parents I have known. Of all the places I have been, the troubles I have seen, and the trials I have endured, let there be no doubt that I continually felt their guidance ahead of me, their support behind me, their love beside me.

2.
SPACE,
THE FINAL FRONTIER

By July 1994, I was three years out of my PhD and I had just joined the Hayden Planetarium as a half-time staff scientist, in a split position with Princeton University. I had been asked at the time to give the keynote address to several hundred supporters of the Astronauts Memorial Foundation (AMF), located at the Kennedy Space Center in Florida. The occasion was a joint celebration of *Apollo 11*'s twenty-fifth anniversary and the dedication of the AMF's newly built Center for Space Education. I was one of a dozen or so members of the AMF Board of Directors, which included some executives from Grumman and Lockheed, some influential Floridians, and an astronaut or two from the Apollo and space shuttle eras.

The local industry and culture live symbiotically with the successes and failures of the space program. Many of the region's hotels, scattered throughout Florida's "space coast," feature the twenty-four-hour NASA cable channel on monitors in the lobbies and in the hotel rooms. The local newspapers routinely report on NASA politics, NASA funding trends, and on the launch of everything from tiny weather satellites to the space shuttle. For my keynote address, I was somewhat apprehensive because that audience's universe was quite different from my own. Historically,

NASA had very little presence in the American Northeast—my birthplace. There are no major research campuses, no launch sites, no museums of space technology.

Space exploration is generally beneficial for the country, for the world, and for the human species. But while my formative years of youth coincided with NASA's Mercury, Gemini, and Apollo programs, they had minuscule influence on my ambitions. I saw who NASA was sending into space. The astronauts were predominantly military pilots representing the various branches of the armed forces. There were no women anywhere in the pipeline, and the chosen men all wore crew cuts at the same time the musical *Hair!*, a celebration of 1960s love and nonviolence, was enjoying 1,750 performances on Broadway. Furthermore, the NASA astronauts seemed to have been selected for their steel nerves and their absence of emotional expression. As far as I could tell, the American agenda was not the exploration of space, but the American *conquest* of space to gain military advantage. This was no secret to well-read adults, but to me it was a slow revelation. After searching for meaningful things to say to this audience of space enthusiasts and educators, I managed to invoke Shoemaker-Levy's impending collision with Jupiter as part of a wake-up call for what could (and perhaps should) sustain a funding trend for the American space program.

In spite of my mild misgivings, I am the first to admit that nothing in this world has the power to inspire forward thinking and visions of the future the way the space program can. But there is a fundamental difference. In the 1960s, the technology of the future was something we all looked forward to. In only a matter of years, we would all be drinking Tang, flitting about in bubble cars and monorails, and visiting Moon bases on holiday. However flighty yesterday's visions of the future were, today the pastime of imagining the reachable future has been lost.

I still remember the day and the moment when the *Apollo 11* astronauts landed on the Moon. I happened to be with my influential childhood friend Phillip Branford visiting his relatives in Virginia that summer. The Moon landing was, of course, one of technology's greatest moments. At ten years of age, however, I found myself somewhat indifferent to the event. It's not that I couldn't appreciate the moment's rightful place in human history, I simply had every reason to believe that trips to the Moon would become a monthly occurrence. The ongoing space program, with each mission more ambitious than the next, served as clear evidence of this future. And then, of course, there was Stanley Kubrick's visionary film *2001: A Space Odyssey*, with its space stations and Moon bases. When you add all this together, voyages to the Moon were simply the next step.

Little did I know that Apollo was to become our last steps out of low-Earth orbit for anyone's foreseeable future. In retrospect, I now regret that I did not feel more emotion back on July 20, 1969. I should have reveled in the landing as the singular achievement that it turned out to be.

In spite of the adventure-romance of the Moon landings and sci-fi films, the funding stream for the space program had been primarily defense driven. Cosmic dreams and the innate human desire to explore the frontier are just not as effective at dislodging $100 billion to go the Moon as a cold war enemy and the mandate of a beloved, assassinated president. Maybe America needs a new enemy empire to resurrect the catalyst for the limitless flow of funds into the defense and space industries.

But suppose we interpret the word "defense" to mean something far more important than what standing armies and arsenals can bring. Suppose defense means not the defense of political borders but the defense of the human species itself. One needn't look far for a fast lesson in survival. When the Shoemaker-Levy 9 Comet slammed into Jupiter's upper atmosphere, it unleashed an equivalent energy of two hundred thousand megatons of TNT on the planet. At ten trillion times the destructive energy of the Hiroshima atomic bomb, this sort of collision, if it happened on Earth, would swiftly render the human species extinct.

If we retain the "defense of the human species" as a mission theme, then we have a genuine cosmic vision to share with today's children. (And it's even better than bubble cars and monorails.) They can be charged with saving life as we know it. We must first acquire a thorough understanding of Earth's climate and ecosystem, which will help to minimize the risk of self-destruction. Second, we must colonize space in as many places as possible, which will proportionally reduce the chance of our annihilation from a collision between Earth and a comet or asteroid—we would then no longer have all our eggs in one basket, as it were.

The fossil record teems with extinct species—species of life that had thrived far longer than the current Earth tenure of *Homo sapiens*. That list includes dinosaurs. The chunks of Shoemaker-Levy 9 were so large, and were moving so fast, that each hit Jupiter with at least the equivalent energy of the dinosaur-killing collision between Earth and an asteroid 65 milion years ago. Whatever damage Jupiter sustained, one thing is for sure: it's got no dinosaurs left.

Dinosaurs are extinct today because they lacked opposable thumbs and the brainpower to build a space program. There would be no greater tragedy in the history of life in the universe than humans becoming extinct—not because we lacked the intelligence to build interplanetary spacecraft, but because the human species itself chose not to fund such a survival plan. The dominant species that replaces us in post-apocalyptic Earth just might wonder why we fared no better than the proverbially pea-brained dinosaurs.

We may still have an opportunity to impart our long-lost visions of the future upon the aspirations of the next generation. I recognize that if you are one of those who have lost all hope in the "youth of today," then you are not alone in either space or time. Adult complaints about degenerate kids tend to cross time and space. Consider the following:

The earth is degenerating these days. Bribery and corruption abound. Children no longer mind parents . . . and it is evident that the end of the world is approaching fast.

Assyrian tablet engraved in 2800 BCE

But if kids were really degenerating from generation to generation, then civilization should have collapsed long ago. So it can't be as bad as people think. We just need creative ways to inspire the next generation.

Shoemaker-Levy 9 and its collision with Jupiter was an omen, just as comets were in ancient times. But this omen was in the form of a shot across spaceship Earth's bow. It's the type of news that can shock us into new paradigms, where our investment in the next generation's dreams of space exploration becomes our species' life insurance policy. Equipped with such a policy, we can be fearless in the face of an asteroid with our name on it.

Unfortunately the "defense of the species" battle cry falls flat in the halls of Congress. In mid-2003 I cosigned an open letter to Congress, along with a dozen other concerned citizens that included Caroline Shoemaker and David Levy (of Shoemaker-Levy 9 fame), as well as physicist Freeman Dyson from the Institute for Advanced Study in Princeton, and Harrison H. Schmitt, the geologist from *Apollo 17*, the last of the manned lunar missions. In the letter, we laid out the entire risk assessment and told what effort would be required for America and the world to do something about it. However real such dangers were to Jupiter, they remain abstractions in the minds of most politicians. But, as a nation, we sure do react to imminent danger with force and resolve. Life in the wake of the September 11, 2001, attacks forced people, Americans in particular, to reevaluate the state of our personal and national security in the air and on the ground, and has led to two expensive wars, one in Afghanistan and one in Iraq. And in the time passed since the space shuttle *Columbia*'s fatal reentry through Earth's atmosphere in February 2003, it seems that everyone has become a NASA critic. After an initial period of shock, followed by genuine mournful expression, no end of journalists, politicians, scientists, technologists, policy makers, and ordinary taxpaying citizens have debated with passion and persuasion, the past, present, and future of America's presence in space.

The two scenarios are linked in an obscure but undeniable way.

Among the complaints about our future in space is the perennial groan that people no longer get excited about the space program. Unlike what has been argued by others, including Dr. Buzz Aldrin, *Apollo 11* astronaut, this lament is not a measure of apathy. It's instead an indicator that space exploration has passed seamlessly into our culture. A nation's culture is what permeates life so thoroughly that its residents no longer take notice. Italians do not stop to notice that their grocery stores contain entire aisles of pasta any more than Americans stop to notice that our stores contain entire aisles of soft drinks, crunchy breakfast cereals, and other products we invented or pioneered. As for the space program, we now pay attention only when something goes wrong. A comparison with the space-faring future world of Stanley Kubrick's *2001: A Space Odyssey*, now fading fast into our past, cannot be avoided. Most observers will complain of our measly earthbound life in the real year 2001. Even though we've got no lunar bases and we haven't sent hibernating astronauts to Jupiter in outsized space ships, we have actually done quite well for ourselves.

The greatest obstacle to the human exploration of space, apart from funding and other earthly political factors, is surviving biologically hostile environments. We need to engineer a version of ourselves, an emissary who can somehow withstand the extremes of temperature, the high-energy radiation, and the meager air supply, yet still conduct a full round of scientific experiments.

We've already invented such things.

They are called robots, and they conduct all of our interplanetary exploration. You don't have to feed robots. They don't need life support. And they won't get upset if you don't bring them back. At any given moment, the typical ensemble of space robots includes probes that, from one year to the next, monitor the Sun, orbit Mars and search for life on its surface, intercept a comet's tail, orbit an asteroid, study Jupiter and its moons, as well as Saturn and its moons. Four of our earlier space probes were given an orbital trajectory with enough energy to escape the solar system altogether, each one carrying encoded information about humans for the intelligent aliens who might recover the hardware.

We have compelling evidence for the existence of barely frozen water on Mars and of liquid water deep within Europa, one of Jupiter's moons. These worlds hold tantalizing prospects for the past or present existence of non-Earth-based life. This news was, of course, beamed to us by semi-intelligent, robotic probes endowed by humans with the capacity to ask and answer many of the questions that humans would ask were we the ones making the trip. We also maintain, at any moment, hundreds of communication satellites as well as a dozen space-based telescopes that see the universe in bands of light from infrared through gamma rays. One of these pass bands, the microwaves, allows us to see evidence of the big bang coming from the edge of the observable universe.

Just because we have no interplanetary colonies, or other unrealized dreamscapes, doesn't mean that our presence in space has not in fact grown exponentially. We should not measure our space-faring era by where footprints have been laid. Nor should we measure it by how many people deify our astronauts or follow the progress of our launches. We should measure our era by how many people take no notice at all. A legacy rises to become culture only when its elements are so common that they no longer attract comment—not because people have lost interest, but because people cannot imagine a world without them.

As for the real year 2001: apart from our flocks of robotic probes, we had a silent ballet of hardware in the heavens. The International Space Station was under construction, just like the one portrayed in *2001* the movie, and it will never know a day without an astronaut on board—our human presence in space is now permanent. The space station was being assembled with parts delivered by reusable, docking space shuttles, each of which have NASA

printed on the side panels instead of Pan Am. Further airline simi-
larities include zero-G toilets with complicated instructions and
platters of unappealing astronaut food.

As far as I can tell, the only things Kubrick's movie have that
we don't have is Johann Strauss's *Blue Danube Waltz* filling the
vacuum of space, and a homicidal mainframe named HAL.

By contrast, space in the 1960s was an exotic frontier traversed
by the few, the brave, the lucky. Every gesture NASA made toward
the heavens caused a spike in the media—the surest evidence that
space was not yet familiar. Back then, as you would expect, many
of us could recite the monikers of the "Mercury Seven" astronauts.
Today, as you would expect, the Columbia Seven became similarly
well known, but only in death. With the space shuttle, America had
launched more astronauts in the eighteen months preceding the
Columbia tragedy than were launched in all Mercury, Gemini, and
Apollo missions during the 1960s.

What does all this mean? The run of space missions from the
1960s, each more ambitious than the one before it, ultimately led to
men walking on the Moon, just as we said we would. Mars was
surely next. These adventures spawned a level of public interest in
science and engineering that was without precedent in American
history, pumping the entire educational pipeline with eager and
inspired students. What followed was a domestic boom in technology that would shape our lives for the rest of the century. A
beautiful story. But, as already noted, let us not fool ourselves into
thinking we went to the Moon because we are pioneers, or discoverers, or adventurers. We went to the Moon because it was the mil-
itaristically expedient thing to do.

Just weeks after the Soviet Union's Yuri Gagarin became the first
person to orbit Earth, President Kennedy addressed a joint session of
Congress on May 25, 1961, and uttered words that still resonate today:

> I believe that this nation should commit itself to achieving the
> goal, before the decade is out, of landing a man on the moon and

returning him safely to the earth. No single space project in this period will be more impressive to mankind, or more important for the long-range exploration of space; and none will be so difficult or expensive to accomplish.

But hardly anyone remembers a separate paragraph, one that preceded the famous one above, in which Kennedy made a powerful appeal to defeat communism:

> If we are to win the battle that is now going on around the world between freedom and tyranny, the dramatic achievements in space which occurred in recent weeks should have made clear to us all, as did *Sputnik* in 1957, the impact of this adventure on the minds of men everywhere who are attempting to make a determination of which road they should take.

This was not, of course, the first time that significant monies were spent on military programs. Kennedy knew, if only implicitly, that while bravery may win battles, science and technology provide security. Science and technology win wars.

From my reading of history, human discovery and exploration have never driven the funding of truly expensive projects, even if our sanitized memories tell us so, and even if the people doing the discoveries are, themselves, discoverers. This fact argues strongly against those who suggest we have not yet walked on Mars because today we have no leaders, or we have lost our drive to explore, or we no longer take risks.

If you want exploration alone to promote spending, consider that any foreseeable mission to Mars will be long and immensely expensive. We are a wealthy nation. We have the money. The needed technology is imaginable. These aren't the issues. Expensive projects take a long time and must be sustained across changeovers in political leadership as well as through downturns in the economy. With no immediate military benefit or economic driver (like space tourism), images of astronauts frolicking on a planet's surface juxtaposed with that of hungry, unemployed factory workers make a powerful anti-funding force.

A review of history's ambitious projects—those that have garnered an uncommonly large fraction of a nation's gross domestic product—demonstrates that only three drivers have been sufficient to create them: *defense* (e.g., the Great Wall of China, the Manhattan project, the Apollo project), the *promise of economic return* (e.g., Columbus's voyages, Magellan's voyages, Tennessee Valley Authority), and the *praise of power* (e.g., pyramids, cathedrals, palaces). For expensive projects that satisfy more than one of these criteria, money flows like rivers. The Eisenhower interstate highway system makes a crisp example: conceived in the post–World War II era to move matériel and personnel for the defense of the nation, yet used heavily by commerce. That's why there's always money for roads.

Low-earth orbit is still a frontier, of sorts. Although today's astronauts are boldly going where hundreds have gone before, the empirical risk of death remains high. With two lost shuttles out of a hundred launches, an astronaut's chances of not coming home are about 2 percent. If those were your chances of death every time you drove your car, you might never drive your car. The *Columbia* astronauts were not unmindful of this risk, yet they took it anyway. They went because the return outweighed the risk itself. I am proud to be part of a species where a subset of its members willingly put their lives at risk to push the boundaries of our existence. They would have been the first to leave the cave and see what was on the other side of the cliff face. They were the first to scale the mountains. They were the first to sail the oceans. They were the first to touch the sky. And they will be the first to land on Mars. But somebody has to write the check. When nobody writes the check, we stall on the last broached frontier.

Rhetoric won't get us there. Nor am I convinced that the fundamentals of human decision making are different today than they have ever been. So, unless space travel becomes so cheap it's not worth a congressional debate, or unless we have investors lined up to sink venture capital into space hotels, or unless we have the reprise of a *Sputnik*-like assault on our national security, as some have predicted will occur with the Far East and their new forays into space, we will simply never go anywhere else.

Actually, there may be a way. But it involves a different shift in what we have traditionally called national defense. If, in fact, science and technology win wars, as the history of military conflict suggests, then, instead of taking count of our smart bombs, perhaps we should be taking count of our smart scientists and engineers. In World War II, those who cracked the German code, who invented radar, and who enabled the Manhattan project were all drawn from their academic labs where they had been conducting curiosity-driven research on the frontier of science and technology.

To attract the most talented students, you need the best projects—not military projects, but pure, curiosity-driven projects. We should search Mars for water, fossils, and life. Liquid water once ran on its surface. No longer. As earthlings who live on a fragile, wet planet, we ought to make this study a high priority. We should visit an asteroid or two and learn how to deflect them. We should drill through the kilometers of ice on Jupiter's moon Europa and explore its subsurface liquid ocean for living organisms. We should explore Pluto and its newly discovered family of orbiting icy bodies in the outer solar system because they contain clues to our planetary origins. We should probe Venus and its atmosphere. Its runaway greenhouse effect tells us that something went horribly wrong. Using people as well as robots, no part of the solar system should sit beyond our reach. No part of the universe should hide from our telescopes, launched into orbit around Earth, the Sun, and elsewhere.

With mission plans and projects such as these, I, as an educator, can guarantee you an academic pipeline stoked with the best and the brightest biologists, chemists, physicists, geologists, astrophysicists, and engineers. And yes, they will collectively form a new kind of silo—one filled with intellectual capital—that will be available when called, just as the nation's best have come when called before.

To die on the frontier without hope that others will follow, because nobody would write the check, is to move backward just by standing still. None of us wants our descendents to reflect fondly on a time passed when America once shined in the timeline of cosmic discovery.

By mid-2001, I had been invited by Pres. George W. Bush to bring some of these collected perspectives on our future in space to the deliberations of a newly formed White House commission. The American aerospace industry had lost more than a half million jobs over the previous two decades, from the consolidation of more than fifty companies down to just five, and from other market forces, including heavy competition from the aero-industries of other nations, especially that of Europe's Airbus. This Commission on the Future of the United States Aerospace Industry contained a dozen commissioners, six appointed by Congress and six appointed by the president himself. For the six appointed by Congress, two were appointed by the majority parties of both houses, one from each of the two minority parties. Since the president was Republican and both Houses of Congress contained a Republican majority, one might expect the process to have forged a heavily partisan commission. But the subject of aerospace and the health of this crucial industry to the American way of life is decidedly not partisan. At no time during the commission's thirteen-month tenure did our deliberations break into partisan politics, in spite of how contentious our conversations got. Two of the commissioners, astronaut Buzz Aldrin and I, brought space expertise to the commission, while everyone else represented the aircraft industry in one way or anther.

The rest of the commissioners and their current or former jobs reads like a who's who of testosterone: The Honorable Robert S. Walker, commission chair and former head of the House Science

Committee; Edward M. Bolen, president of the General Aviation Manufacturers Association; R. Thomas Buffenbarger, the president of the International Association of Machinists and Aerospace Workers; John Douglass, brigadier general for the air force (ret.), former assistant secretary of the navy, and president and CEO of the Aerospace Industries Association; F. Whitten Peters, former secretary of the air force; John Hamre, former deputy secretary of Defense; William Schneider, chair of the Defense Science Board; and Robert J. Stevens, president and COO of Lockheed Martin Corporation and now its CEO. Even the commission's two women fit the testosterone criterion: Heidi Wood, executive director of Morgan Stanley and senior analyst for aerospace, defense, and defense electronics; and Tillie K. Fowler, former member of Congress and of the House Armed Services Committee.

Our first meeting was held in Washington, DC, and took place on October 2, 2001, just three weeks after the September 11 terrorist attacks. The events of September 11 magnified the importance of the commission's agenda: to recommend to the White House, Congress, and other relevant government agencies what strategies should be implemented to assist, or to rebuild, a failing industry—an industry that enabled a way of life and a level of security we have all taken for granted in America's post–World War II era. But I felt especially violated because the World Trade Center's twin towers sat a mere four blocks from my living room window in lower Manhattan. My family lives in a converted loft within full view of city hall and City Hall Park, at the intersection of Broadway and Park Row. The "Canyon of Heroes" parade route ends at this spot, where, in what seems to be an annual ritual, defenestrated paper strips—hundreds of tons of the stuff—flutters onto the New York Yankees motorcade after they win the World Series. This same parade route honored John Glenn in 1962 upon return from orbit in his Mercury capsule *Friendship 7,* and upon his return from space shuttle flight STS-95, thirty-six years later.

But on September 11, 2001, the social, emotional, and political perspectives that I carry through life were forever altered. While I reflected on these events, my anger and frustration remain strong, if unfocused. Bearing witness to what makes world headlines and what carries a nation into battle is a heavy burden. My family spent the following twelve days just north of the city as war refugees from lower Manhattan. For each of the first ten days, I slept fourteen hours—two and a half times my nightly average. For the next several days, I spent most of my waking hours in stunned silence. For the two months that followed, the sound of sirens (normally, a form of acoustic wallpaper to a city dweller) rustled my nerves, which had been continually immersed in the sounds of rescue vehicles for nearly two hours—until the collapse of the south tower, when the air went morbidly silent.

For the two months that followed, the mere sight of Park Avenue South, the route I walked while pushing my nine-month-old son in his stroller and carrying my five-year-old daughter, would cause my muscles to twitch, in silent, yet autonomic remembrance of my expended energy escaping three miles to Grand Central Terminal, northward to my parents' quiet home in Westchester. I suppose these were all symptoms of a form of shell shock that were slow to fade.

I remain fearless of airplanes. But during a trip to Los Angeles on a Boeing 767, I couldn't keep my mind from drifting: What's the largest piece of this airplane that could crash into the World Trade Center, explode out the other side, and survive intact? The landing gear? My computer battery? My belt buckle? My wedding ring? How quickly would I die? One second? A tenth of a second? As a college wrestler and as an amateur martial artist, how many terrorists could I restrain?

I remain different in several ways. My emotional mind has been somewhat separated from my rational mind. They were formerly married to each other, in careful balance, but with the prenuptial agreement that my emotions would never override decisions that required the benefit of rational thought. From fire to ice and back, feelings went unchecked. The endless parade of tourists trudging past my living room window, bestrapped by cameras, to gawk at the still-smoldering remains of America's worst disaster, made me irrationally angry. "Which way is Ground Zero?" they would ask, while covering their noses and mouths, not wanting to breathe the clouds of Ground Zero smoke that I breathed every day. Although finance defines the region, fifty thousand people continue to claim downtown Manhattan as their neighborhood. I was not alone in my sentiments.

My sister, Lynn, hasn't lived in the city in nearly two decades, but she knew and loved the World Trade Center from the time she worked for the city as a mounted urban park ranger, and she gave scheduled tours of the area's parks and monuments. During her first visit to my home after September 11, she looked out the window and made the arresting comment, "I find it easier to believe the towers were never there, than to believe that they are now gone."

The picture-taking tourists were generally respectful of the makeshift shrines along the neighborhood streets, near churches, police stations, and fire stations. Their touristy chatter fell silent by the force of reverence as they passed the vistas to the tangled wreckage. Of course, what had *I* done my first day back in the neighborhood? I walked the route and quietly took pictures. Realizing the hypocrisy, my rational mind slowly rose up to be heard. After a week, or so, when asked, I started directing tourists to the best views of the wreckage. I did it because it was the right thing to do. Ground Zero belongs to America. Ground Zero belongs to the world. Ground Zero is the hallowed graveyard for nearly three thousand souls. It just happens to lie in my backyard.

Other changes surfaced within me that remain to this day. I leave work a little earlier. I hug my children more often. I am more likely to talk to strangers. I am more easily saddened by sad events. And, as is true for so many, I have become intolerant of intolerance. The NYPD changed too. They became genuinely helpful and friendly. They even smile and pose for pictures with passersby. An extraordinary sight to a New Yorker.

Our local fire department, one and a half blocks away, lost six men. They lost "only" six because they were the first on the scene, assisting escapees from the north tower, the first one hit. Rescue workers who came later, after the south tower was hit, went there. But the south tower collapsed first, one acre of offices per floor, 110 floors, burying all near its base. Fire stations farther away in Manhattan, with longer travel times to the site, lost upward of a dozen men. For nearly two years, the sidewalks outside these fire stations spilled forth with candles and flowers. Another shrine had lay along the Hudson River, near a pier that contained a morgue and a makeshift forensic lab that identified the remains of the dead being shuttled from Ground Zero. You couldn't walk more than a half dozen blocks anywhere in the city without an encounter with one of these quiet reminders that something very, very bad had happened.

On the morning of September 12, I sent a descriptive e-mail to a small circle of family, friends, and colleagues of my previous day's escape from lower Manhattan. Within hours, that message became widely redistributed via people's e-mail address books. Among the thousands of responses I received came from a man who mailed two cuddly stuffed animals. He did this after reading of my daughter's sadness that her stuffed animals would have dust all over them and that we would not soon return to the apartment. Sometimes little gestures are big gestures.

At less than one year old, my son was too young to know or remember anything that happened in those days. He still cried when he was hungry and laughed at peek-a-boo. My daughter occasionally talked about the tragedy, but in a way that leads me to believe she's just fine. "Daddy, if the bad men on the airplane are dead, how did the newspapers get a picture of them?" "Daddy, if the World Trade Center were across the street, where the city hall fountain is, then the people who fell from the windows might have fallen in the water and lived." "Daddy, even though the World Trade Center is gone, the World Financial Center is still there. Maybe when they clear away the dust, we can go back to its park and play."

By two weeks after September 11, we had moved back to our downtown loft, but only after an eighth of an inch of World Trade Center dust was cleared from every horizontal and vertical surface in our apartment. This was a four-day job, with two of those days employing six people wielding brooms, microfiber sponges, and HEPA vacuums. This dust layer, a mixture of pulverized concrete, wallboard, other silicates, and traces of asbestos, had flowed through the panes of our closed windows. The dust cloud of the collapsed towers had been so thick and dense that many of our neighbors, those who had left their windows open on that beautiful end-of-summer day, did not move back for months. At least one of our neighbors had to discard every drape, every sheet, and every item of clothing that was left behind. Some never returned at all.

While the media and Congress slowly marshaled feelings of anger and patriotism, I had no such luxury of thought. We were just trying to conduct our lives in what was effectively a war zone. Military vehicles had blocked most of our local streets. The power line that supplied the New York Stock Exchange had police guards at every node. This line had been swiftly laid, above ground, enabling the exchange to open just one week after September 11. My daughter's elementary school, P.S. 234, just three blocks north of the north tower, had closed to make room for the rescue and recovery efforts of Ground Zero personnel. She would occupy two temporary school shelters before returning six months later, only after all fires had been extinguished. Depending on which way the wind blew, the Ground Zero fires brought a dusty, smoky stench—the vaporized blood of the towers—to all of lower Manhattan. For every hunk of metal extricated from the wreckage by the cranes and tractors, fires would break out below. At night, with the brilliant construction lights illuminating the site, you could see from our dining room window plumes of smoke rising fifty stories high. Of course, the smoke rises right where the towers used to stand in view. Sanitation trucks stopped sweeping, and instead they sprayed water up and down the streets, keeping the kicked-up dust to a minimum. Meanwhile, large, flatbed dump trucks hauled away ton after ton of Ground Zero debris, twenty-four hours per day. At home, we purchased and continuously ran two high-volume HEPA air filters, cycling the apartment's air four times per hour.

Before we moved back, I collected dust samples from our windowpane and brought them to a midtown lab for analysis. The asbestos content fell below the measurement criteria for it, allowing the full cleanup without donning moon suits. Within the dust sample I also noticed teeny flakes of black carbon. Probably charred office paper. But I kept wondering if the windblown remains of immolated victims in the precollapse fires comprised some percent of this. The fires created a furnace hot enough to render molten the steel cores of the World Trade Center towers. Before my apartment received professional cleaning, I collected a vial's worth to keep as a kind of reliquary—in remembrance of a tragic portal through which we had all passed.

September 11 itself began as one of those perfect fall days: seventy degrees, and not a cloud in the sky, and with the crisp visibility that comes with no haze and extremely low humidity. I happened to be working at home that day. My wife went to work at 8:20 AM. I left at the same time to vote in NYC's mayoral primary. My son, nine months old at the time, was at home with our nanny. My five-year-old daughter was attending her second day of kindergarten. Lineup time in the yard was 8:40 AM in full view of WTC-1, the north tower.

When the first plane hits at 8:50, they evacuate the school without incident. I notice WTC-1 on fire in a high floor upon returning from voting, about 8:55 AM. Large crowds of onlookers gather along the base of City Hall Park as countless fire engines, police cars, and ambulances scream past.

I run home, grab my camcorder, go back out to the street, and film the events before me. My camera's high-power zoom enables me to judge the ensuing damage to the tower so that I make an informed decision about if and when I would need to retrieve my daughter from school. Through the 20:1 zoom I see not just some flames coming out of some windows, but four or five entire floors ablaze, with smoke penetrating floors still higher.

This was upsetting enough, but then, among the papers and melted steel fragments fluttering to the ground, I notice that some debris falls quite differently. These aren't parts of the building. These are bodies. Human bodies. Jumping from the eightieth floor and tumbling in a surreal, slow-motion fall to their death below. I notice about ten such falls, and quickly remember that a falling human reaches its highest velocity through Earth's atmosphere at about two hundred miles per hour. After that, the slowly parting air resists further acceleration. From the World Trade Center, or from any sufficiently tall building, you reach this terminal velocity after a fall of only about twenty floors, yet you keep falling. So the slow-motion falls I had noticed was my brain's commentary that the bodies had stopped accelerating past the sixtieth floor, descending to the pavement below thereafter at a constant speed.

Then, a fiery explosion bursts forth from the northeast corner of WTC-2, about two-thirds of the way up, at least the sixtieth floor. The fireball creates an intense impulse of radiative heat from which we all cower. From my vantage point, I could not see the second plane, which had hit the tower's far side. I didn't know at the time that a plane caused the explosion. I first think a bomb did it, but a bomb's telltale shockwave, which rattles and often shatters nearby windows, does not accompany this explosion. All I feel and hear is a low-frequency rumble.

WTC-2's fireball extends all the way across to WTC-1, nearly one hundred yards away. The flames are followed by countless thousands of sheets of paper, cast forth as though from a cannon, as they flutter to the ground, falling within Broadway's Canyon of Heroes, but this time on the bodies of those who had escaped immolation by jumping to their deaths.

That the second tower was now on fire made it clear to us all on the street that the first fire was no accident and that the WTC complex was under terrorist attack. After capturing the explosion on tape and the sounds from the horrified crowd surrounding me, I stop filming, and go back inside my apartment.

How much more upsetting can this get? As more and more and more and more and more emergency vehicles descended on the World Trade Center, I hear a second explosion in WTC-2, then a volcanically loud, low-frequency rumble that precipitates the unthinkable—a collapse of all the floors above the point of explosion. First the top surface, containing the helipad, tips sideways in full view. Then the upper floors fall straight down in a demolition-style implosion, taking all lower floors with it, even those below the point of the explosion. A dense, thick dust cloud rises up in its place, which rapidly pours through the warren of streets that cross lower Manhattan.

I swiftly close all our windows and blinds. As the dust cloud engulfs my building, an eerie darkness surrounds us—the kind that falls just before a severe thunderstorm or tornado. I look out the window and can see no more than about a foot away. But within that foot, I see the nearest turbulent eddies of the rolling cloud, made visible by the tumbling of sheets of unburned office paper, countless millions of them, cast forth from the collapse of all unburned floors.

During the next fifteen minutes, visibility slowly grows to about one hundred yards, and I see about an inch of white dust everywhere outside my window: on the sills, the sidewalks, the streets, and on the leaves of the trees of City Hall Park, leaving the region to look as though it had just endured an early autumn snowfall. At this moment I realize that every single rescue vehicle that had parked itself at the base of the World Trade Center is now buried under 110 collapsed floors of tangled metal and concrete, mixed with multiple feet of dust. This collapse takes out the entire first round of rescue efforts, including what are hundreds of police officers, firefighters, and medics.

As visibility further increases, I look up. There was blue sky where WTC-2 used to be.

With full communication between me and my wife, established via intermittent connectivity between our two different cell-phone carriers and two different land-line carriers, I decide it's time to get my daughter, who, I have learned, was taken by the parents of a friend of hers to a small office building, six blocks farther from the WTC than my apartment. While I dress for survival in boots, flashlight, wet towels, swimming goggles, bicycle helmet, and gloves, I hear another explosion followed by a now-familiar volcanic rumble that signaled the collapse of WTC-1, the first of the two towers to have been hit. I see the building's iconic television antenna descend straight down in an implosion twinning the first.

The ensuing dust cloud was darker, thicker, and faster moving than the first. When this round of dust reached my apartment, fifteen seconds after the collapse, the sky turned dark as night, with visibility of no more than about an inch. I foresee no hope of survival for any of the rescue personnel who were on the scene.

The cloud settles once again, leaving about three inches of dust outside my window. A dark cloud of smoke now occupies the spot where two 110-story buildings once stood. This cloud, however, was not the settling kind. It was smoke from ground-level fires. By now, the loft's air is getting harder and harder to breathe, which forces my decision to evacuate—especially with the likelihood of underground gas leaks from severed and smashed pipes. I would not get my daughter and bring her back home. I would get my daughter and escape somewhere. So I load up my largest backpack with longer-term survival items such as cans of tuna, a knife, bottled water, diapers, and baby food. I put my infant son in our most nimble stroller and leave with our nanny, who then walks across the Brooklyn Bridge toward her home.

I go to where my daughter is held, upwind from all debris, on a quiet street. She, along with other kids brought there, is in good spirits, but clearly shaken. I still have a crayon drawing of hers, sketched while waiting for me to arrive, which shows the towers with smoke and fire coming from them, as only a five-year-old child could draw. "Daddy, why do you think the pilot drove his plane into the World Trade Center?" "Daddy, I wish this was all just a dream." "Daddy, if we can't return home tonight because of all the smoke, will my stuffed animals be okay?"

From the calm of an upholstered couch in the office where my daughter was kept, with my son under one arm and my daughter under the other, I realize that, fully loaded, each tower holds ten thousand people. At the time, I had no reason to believe any of them survived. Beneath the towers sprawls a veritable city comprising six subterraneous levels containing scores of subway platforms, plus a hundred or so shops and restaurants. The towers simply collapsed into this hole—a hole large enough to have supplied the landfill for the World Financial Center across the highway from the World Trade Center.

I reconnect with my wife by 4 PM, meeting her just north of Union Square Park. We hike another mile north to Grand Central Terminal for our ride to the safety of my parents' home in Westchester, twenty-five miles north of Ground Zero.

The twin towers' actual death toll of about three thousand was much less than I had imagined, but nonetheless exceeded that of Pearl Harbor, and was more spectacularly tragic than the *Titanic*, the *Hindenburg*, and Oklahoma City combined. I am no longer the same person I once was. I suppose that my generation now joins the ranks of those who lived through unspeakable horror and survived to tell about it. How naive I was to believe that the world is fundamentally different from that of our ancestors, whose lives were changed by bearing witness to the twentieth century's vilest acts of war.

Imagine carrying all this emotional and intellectual baggage into the first meeting of the president's aerospace commission. I was ready to change the world. But of course, so was everyone else on the commission. The difference was that if I were to change the world, it could be through only whatever power of persuasion my academic pedigree could engender. For my fellow commissioners to change the world, it would be through the power they wielded directly. This distinction left me in a peculiar quantum state. Everyone else on the commission represented some kind of a constituency whose views they carried into our deliberations. This simple fact establishes boundary conditions to what any one of them could say, lest they offend their supporters or compromise the political stand with which they are associated. So while I wielded no power over agency, I was free to say whatever I felt, without fear or concern about my future in Washington, DC, which falls outside of my career trajectory. I later learned that the White House staffing office had sought precisely what I had brought to the commission.

As an example of the company I kept for thirteen months across two dozen meetings and public hearings, and two world tours, one to Europe and Russia and one to China and Japan, the first phrase of the first sentence I heard from Commissioner John Douglass was the following, "When I was U.S. Military Representative to NATO in Europe during the Gulf War. . . ." We had been discussing the emergent threat to Americans abroad after September 11, and we were comparing security stories. Douglass went on to describe three separate kidnapping attempts on his life and the different reactions of his bodyguards to these assaults. Others followed with equally extraordinary stories of military or security encounters, and clever things for pilots to do the next time they get hijacked. My favorite one involves a latter-day autopilot feature where the plane, on the push of a button, automatically takes permanent control of all panels, navigating via GPS without a pilot, and landing safely at the nearest airport, all by itself. Another solution has the pilot take the plane in a tight barrel roll. No way for anybody to stand up straight. The terrorists get dizzy, everyone vomits, and you fly the plane home.

Not much I can add or subtract here. Except the dinner that followed our first meeting in Washington was not held at "Vegetarian Valley." We ate at Ruth's Chris steak house, of course.

My service on the Aerospace Commission ended with a 198-page report titled "Anyone, Anything, Anywhere, Anytime" that we hand delivered to Vice President Dick Cheney in the East Wing of the White House in November 2003. The report's introduction contained a strongly worded appeal for the United States to recognize the value of the aerospace industry to our security and to the health of the nation's economy. It also identified a renewed investment in space exploration—one that contains a vision where we actually go somewhere—as the most fertile means to attract the next generation of scientists and engineers into the workforce. These are the people who make tomorrow come. Among the chapters that specifically referenced the *aero* part of aerospace, one was devoted entirely to our future in space.

One never knows how much impact commission reports really have on prevailing thought or political policy. Given how many commissions get established for one reason or another, by one government agency or another, at times it seems that commissions are just a means for policymakers to appear as though they are working on a problem whether they are or not. At best, members of Congress might wave the document in the halls of Congress as a recommended guide to debates and legislation. At worst, the report gets shelved and forgotten. A middle ground exists, however, one where an otherwise barren or hostile political landscape becomes a place where ideas and visions can take root.

Not three months after our final report, space shuttle *Columbia* was lost on reentering Earth's atmosphere after a structural damage in the protective tiles had left the orbiter susceptible to overheating and eventual breakup. I happened to witness the launch of that shuttle mission, live from the viewing stands of Cape Canaveral. I captured on my camcorder the jubilation of the astronaut families during the shuttle's thunderous ascent. Included was the flag-waving wife of Ilan Ramon, the first Israeli astronaut. This launch was the first (and only) of any kind I have ever attended, and so has left a singular imprint on my awareness of the ambitions, the drama, and the emotions contained within the manned space program.

With the space shuttle fragments and the astronauts' remains scattered across Texas, you might think the bereaved families would cry out against our presence in space. But something different happened. The astronauts' loved ones declared unanimously that the adventure must continue. And with the rest of us knowing in our hearts that if the lost astronauts could somehow be questioned, they would agree with this sentiment, the country (via editorials, op-ed pages, letters to the editor, and TV and radio talk shows) declared that if we were to put human lives at risk, it should be to accomplish a vision greater than simply going to and from low-earth orbit. Humans have not ventured farther from Earth than the distance Washington, DC, is from Boston since our last trip to the Moon in 1972 onboard *Apollo 17*. With the Aerospace Commission report serving as a substrate, the ensuing commentary and public sentiment flourished. Add that China had just launched its first taikonaut, Yang Liwei, into Earth orbit, and that the *Beijing Morning Post* quoted Ouyang Ziyuan, its chief space scientist, as saying, "Our long-term goal is to set up a base on the Moon and mine its riches for the benefit of humanity" and you have all the ingredients to trigger a Kennedy-like space vision for America.

Only a year later, and only fourteen months after the commission report was filed, President Bush delivered a speech at NASA's Washington, DC, headquarters declaring the time has come to embark on a new era of space exploration—one that will take humans back to the Moon, onto Mars, and beyond. Science would be a centerpiece of this vision. Such a vision will require that the president appoint a commission to study how this plan can be made to succeed in the fickle climate of Washington politics and funding cycles.

Sure enough, within days of the president's speech, I received the call from the White House inquiring about my interest in serving on yet another commission. This time, however, the entire subject was space. It would be the president's nine-member commission on the "Implementation of the United States Space Exploration Policy." Once again, I agreed. Just as September 11, 2001, had stirred me to serve with heightened purpose on the Aerospace Commission, so too did the loss of the *Columbia* space shuttle, after having attended its launch, create within me an irreversible and irresistible sense of duty.

Unlike the Aerospace Commission, where I was the only scientist (and educator), this time there are four: two planetary geologists, Paul Spudis, a Moon expert from the Johns Hopkins University's Applied Physics Laboratory, and Maria Zuber, a Mars expert and chair of MIT's Department of Earth and Planetary Sciences; a planetary geochemist, Laurie Leshin, specialist in the formation and evolution of the solar system, and director of Arizona State University's Center for Meteorite Studies; and an astrophysicist (me). Others on the commission include Carly Fiorina, the CEO of Hewlett Packard; Lester Lyles, air force general (ret.), and former commander of the Air Force Materiel Command; Michael P. Jackson, former US Department of Transportation Deputy Secretary; Robert S. Walker, chairman and CEO of the Wexler & Walker Public Policy Associates; and Pete Aldridge, chair of the commission, with forty-five years in the aerospace sector, most recently serving as Under Secretary of Defense for Acquisition, Technology, and Logistics. Each one of us carries a specifically tuned expertise to address the presidential charter of the commission, one of which included exploiting in situ resources on the Moon, Mars, and elsewhere in whatever way may be required to realize the vision.

Who knows what the near and long-term effect of this commission's final report will be? But we can be confident that America's finest hours are yet to come.

3.
SCIENTIFIC ADVENTURES

I 've worked hard to expose my daughter to the laws of physics. Actually, she performs most of the experiments herself. She once dropped twenty-three overcooked peas, one by one, from her dinner plate to the ground. This particular experiment highlighted the conversion of gravitational potential energy to kinetic energy (the peas gain speed as they fall), and the nature of inelastic collisions (the peas flatten, instead of bounce, as they hit the floor). During her experiments in fluid dynamics, she poured a cup of apple juice onto her dinner plate, and then poured it back into her cup. She repeated these actions until all the juice had spilled onto the dinner table. She then watched the puddle drizzle down through the seam in the table's leaves to become a puddle on the floor. After dinner, she climbed down from her booster seat and stepped into the puddle, scattering the juice all over. I love it! And yes, I am the one who cleans up after her.

So much of what shapes and comprises what we call "common sense" derives from a careful assessment of the way the world works. So horror descends upon my scientific soul every time my daughter requests to see the Walt Disney classic *Mary Poppins*. The first time I inserted the video into our VCR (remarkably, I had never before seen the film), I did not know what to expect. What unfolded before my eyes thoroughly and purposefully violated nearly all known laws of physics. In Mary Poppins's first appearance in the film, she floats while holding her outstretched umbrella. Okay, maybe that could happen. But upon entering the children's house (after she gets the offer to be the live-in nanny) Mary Poppins slides *up* the banister and precedes to remove from her ten-inch handbag all manner of oversized room trimmings and furniture to make her room and term of employment a bit more comfortable. Later, Mary has a conversation along a London sidewalk with a dog, where she speaks English and the dog speaks dog. Shortly thereafter, while at Uncle Albert's house, Mary serves tea to all assembled while they are laughing and afloat near the ceiling. While frolicking upon London rooftops, after having been sucked up a chimney, Mary Poppins creates stairs through the air out of ascending chimney smoke that bridges from one building to the next. She then leads a procession across it. And at the end of the film Mary has a conversation with the bird head that forms the end of her umbrella handle.

Not long ago, Mary Poppins would surely have been burned at the stake for being a witch. Now she is a cherished Disney icon. My daughter must now reconcile her own nascent common sense with a story that mocks the laws of physics. Perhaps I shouldn't single out *Mary Poppins*. *Alice's Adventures in Wonderland* is no better— except that Wonderland offers no pretense of being in downtown London. The same goes for Never-Never Land in *Peter Pan*, which is off behind a star somewhere,* and Munchkin Land in the *Wizard of Oz*, which, we are reminded, is not in Kansas.

At the risk of sounding like a curmudgeon, allow me to say that one of society's greatest ills is the astonishing breadth and depth of its scientific (and mathematical) illiteracy. Just listen to the circuitous reasoning that some people invoke to justify why they should not wear seatbelts while driving: They are too restrictive; they are too uncomfortable; seatbelts are for sissies. After their

---

*The precise directions are: "Second star to the right, straight on until morning."

explanation is given, ask them whether they have ever taken a high-school or college-level physics class. The answer will be no. College physics is where you learn about inertia and see demonstrations of Isaac Newton's famous law: "Things in motion tend to stay in motion unless acted upon by an outside force." In a curious revenge of physics laws, while not all taxis will stop to pick me up on the street corner in favor of a white person farther down the block, many of these same drivers don't buckle their seat belts either.

No, I do not blame all science illiteracy on Disney, or on Hollywood, as is customary for other problems of society. But I do blame it on how cavalier society treats skills that promote critical thinking—the kind of thinking that enables you to use the scientific laws of nature to judge whether someone else is a crackpot. Children in kindergarten and elementary school routinely take art classes that promote creativity instead of taking classes that explore how nature works. Just look at the horizontal and vertical display surfaces in the homes of parents with young children—refrigerator doors included. The pasta collages will far outnumber the science experiments. Children are also encouraged to read fantasies and fairy tales in which, as best as I can tell, there are no laws of nature at work. Yet we all sit back and wonder how cults can form, how billions of dollars per year can be spent on astrologers and psychics, and how innocent people can be bilked of their savings by paranormal swindlers.

In a recent story in the *New York Times*,* the headline read "A Police Sting Cracks Down on Fortune-telling Fraud." The article recounts several cases of people who were convinced by a fortune-teller that they were cursed or otherwise diseased, requiring a cure, multiple visits, questionable herbal treatments, and large sums of money. One woman in particular had trouble sleeping and, to her credit, first went to physicians, psychiatrists, and priests. But when none of them could offer help, she visited a fortune-teller who diagnosed her as having "a lot of negativity in her aura." The fortune-

---

New York Times, June 16, 1999, metro section, p. B1.

teller prescribed a root remedy that required a trip to the Middle East to obtain. Of course it was the fortune-teller who made the trip—at the woman's expense. Three thousand dollars later, the woman suspected fraud and went to the police. What strikes me hardest about this story is that the woman, who owns an insurance agency, is quoted as saying, "I am not naive or unintelligent."

I don't know how many critical-thinking skills are required to be a good insurance agent. You might hope there's a few. They insure you, your loved ones, and your property, so perhaps some math and logic would be involved in this effort. She exhibited no critical-thinking skills.

Here are a few things the woman did *not* say:

"Gee, if I had been more skeptical of the fortune-teller then I wouldn't have gotten robbed."

"I had a temporary lapse of judgment but that won't happen again."

"Society has duped me into thinking I am intelligent even though I have hardly any capacity to evaluate the statements and claims of others."

Suppose the woman were a lawyer instead of an insurance agent? Suppose it was her turn to be a juror? What would then happen within the legal system? I cannot speak for the federal courts, but I gleaned some insight to these questions during my first stint on jury duty in Manhattan's county court. Having never, until recently, spent more than several years in the same place, or even the same municipality, during my adult life, I had never been called for jury duty, which typically requires a minimum duration of residency. All I knew of courtroom drama was what I watched on prime-time television, featuring eloquent lawyers and swayable jurors. When I was finally called to serve, in November 1997, I had been a resident of Manhattan for three years. When the time arrived, I went willingly and patriotically. I even got dressed up in my best academic tweed. In anticipation of a long wait in the waiting room, I brought my laptop and newspapers to read.

About fifty of us were there, some of whom looked impatient and haggard. They were in their third and final day of waiting. Others, like me, were freshly pressed and wide eyed. The waiting room happened to have a running television perched in the corner, but it was mounted so high that nobody could reach it to change the channel. And there was no telling where the remote was. I don't watch much daytime television, so I cannot distinguish the normal from the unusual. But on that day, and at that time, there was a Jerry Springer marathon of four consecutive hours. I had never before seen Jerry Springer's talk show and I knew nothing about its interview philosophies or its choice of guests.

We were all there, pretending to do important work at the tables and on the couches of the waiting room. And we were all trying to ignore the television when two of the talk-show guests broke out into a fistfight. Our eyes were transfixed and our mouths were agape. I assumed the fight to be a rare moment. Nope. The next set of guests also broke into a fight. I forgot why. Maybe it was the one where somebody's transvestite boyfriend had a secret love affair with the girlfriend's father. We all sat there and watched guest after guest, fight after fight, episode after episode. And we shamelessly cheered the emotional outbursts of each guest who was wronged. By early afternoon, I was finally called for the juror selection process, but not without having borne witness to the most lawless show on television in the hallowed halls of New York City's criminal courthouse.

After a shorter wait outside an actual courtroom, the presiding judge invited a group of us inside for attorney questioning. As others went before me, I was fascinated by the questions and the answers—all attempting to probe whatever biases we might have toward the defendant, who was present and in full view with his lawyer. What might they ask me? What biases might I have? One thing is for sure, they were hell-bent on probing everyone's livelihood. At the time, I happened to be co-teaching (via guest lectures) a possibly relevant seminar at Princeton University. The questioning attorney began:

What is your profession?
Astrophysicist.
What is an astrophysicist?
An astrophysicist studies the universe and the laws of physics that describe and predict its behavior.
What sorts of things do you do?
Research, teach, administrate.
What courses do you teach?
This semester I happen to be teaching a seminar at Princeton University on the critical evaluation of scientific evidence and the relative unreliability of human testimony.
No further questions, your honor.

I was on my way home twenty minutes later.

I suppose I should have been happy to be dismissed. It meant I could go back to work, or go back home and spend time with my family. But I was upset. Not for myself but for our legal system that eschews rational thought. It became easy for me to understand how O. J. Simpson, and the police who beat Rodney King, could be acquitted in the face of strong evidence against them. Emotional truths woven by lawyers in the court of law are apparently more important than the truths of actual events. I have grown to fear for those whose innocence became trapped within the legal system.

From what I know of courts of law, during the questioning of witnesses, yes/no and multiple-choice questions are common. But the laws of physics do not lend themselves to such responses without incurring a major misrepresentation of reality. In my first year as a staff scientist at the Hayden Planetarium, I was called by a lawyer who wanted to know what time the sun set on the date of a particular car accident at a particular location. This question can be answered precisely, but later in the conversation I learned what that lawyer really wanted to know: what time it got dark outside. He was going to compare the time of sunset with the time of the car accident, and had been assuming that everything gets dark the instant the Sun dips below the horizon. His question was poorly formed for the information he was seeking. A better question might have been, "What time do the dark-sensitive streetlights turn on? But even for that question, the presence or absence of clouds and the shadows of nearby buildings can affect the "right" answer.

Although I was tainted goods in the jury selection box, on another occasion, I managed to help convict a person who was charged with a fatal hit-and-run accident. The driver of the vehicle had a photograph of himself, claiming it was taken at the time of the incident and that he was nowhere near the scene of the crime. The defense attorney asked me if I could verify the claimed time of the image from the lengths of shadows laid by cars and people in the photo. I said sure. If the exact date and location of the crime are known, then there exists only one time of day for which the Sun will create a shadow of a given length in a given direction of a given object. Armed with some handy software on the Sun, Moon, and planets, I made simple measurements of the shadows within the photograph. I provided the lawyer with the time of the photo, plus or minus twelve minutes. The suspect's alibi was off by several hours. I suppose he never knew what I had known since age fourteen. In the courthouse of the universe, the laws of physics do not lie, nor are they influenced by anybody's emotional state, and they apply equally to everyone.   .

When scientists invoke the scientific method, our ways are not as mysterious or as foreign as you might presume. The scientific method forces the researcher to go to any extreme necessary to minimize bias during the acquisition and interpretation of data. The biggest source of error and bias in the acquisition of data happens to be the person who conducts the experiment. A researcher's mood, attitude, political leanings, bigotry, and prejudice have all influenced the integrity of scientific data in the past. The most famous deluded experimenter in the history of astronomy was Percival Lowell. In his studies of the planet Mars he "saw" networks of canals connecting areas of polar water supplies to vegetation and cities. The entire public-works project was presumed to be built by intelligent Martians. Lowell drew detailed maps of what he saw and triggered an entire generation of fantasies about life in the universe. This episode would be laughable were it not for Lowell's otherwise distinguished reputation as a first-rate astronomer, best known for launching the systematic search for planet X, which led to the discovery of Pluto. Without chart recorders, photographs, or other means to acquire data, the severe shortcomings of human senses were readily revealed.

Why then, in the court of law, is eyewitness testimony among the most coveted forms of evidence? One or more eyewitnesses can send you to your death. At least nobody ever died from biased data in astrophysics.

Sometimes ignorance of the laws of physics can have innocent, and even playful, consequences. From eleventh grade through the middle of graduate school, I invested my principal athletic energies in the sport of wrestling. I accomplished this not to the exclusion of training my mind, but I was nonetheless serious about my athletic commitment for reasons (I would later learn) that had less to do with my personal athletic ambitions than societal expectations. I was captain of my high school's team and wrestled varsity in college at the 190-pound-weight class, where there was good incentive to not gain a pound because the next-higher category was "unlimited."

Many sports such as rowing, swimming, and cross-country skiing represent extreme physical challenges. But if you have ever wrestled, you will say that wrestling is the most taxing sport you have ever attempted. All you need to do is turn your opponent so that his back lays against the wrestling mat for about one second. Then you win. An entire match lasts eight minutes. To be good at it, all your muscles must be strong, especially those of the upper body.

You must also be flexible, quick, and have near-infinite physical stamina. Lastly, you must intuitively understand vector diagrams from the laws of physics. Knowing balance points, tipping points, strength points, weak points, center of mass, and leverage points are all factors in moving your opponent to his back. I qualified on most counts, although I was almost too flexible. I was quicker than practically all my opponents. And I certainly knew my force diagrams. My average opponent, however, was four inches shorter. Since we weighed the same (190 pounds), with minimal body fat, my opponent's muscles were therefore always larger. I was consistently the weaker wrestler since biophysics dictates that muscle strength is proportional to muscle cross section. My task was to stay clear of vise-grip muscle holds and to keep my center of mass from getting too high relative to my opponent. My opponent's task was to tame and outmaneuver my long and unwieldy limbs.

In the early 1990s, after completing my PhD from Columbia University, I was appointed as a postdoctoral research associate at Princeton University's Department of Astrophysical Sciences, which was all somewhat later than my wrestling prime. I would nonetheless occasionally roll around with the varsity team. In the third year of my three-year postdoctoral appointment, PBS filmed me for a multipart series titled *Breakthrough: The Changing Face of Science in America*, which profiled a dozen or so active scientists from underrepresented ethnic backgrounds. I was featured for seventeen minutes in the one-hour episode titled "Path of Most Resistance," the title of which was selected by the producers from a line in my 1991 PhD convocation speech. My episode profiled two physicists and two astrophysicists. Part of the program's intent was to give the viewer a full picture of the life and times of the featured scientists. In my case the producers showed some gratuitously embarrassing baby pictures and home movies from my childhood. The producers and film crew also trailed me to the Andes Mountains of South America, where they documented my multinight observing session on telescopes at the Cerro Tololo Inter-American Observatory.

Back on campus, they also wanted footage of me working out with the Princeton University wrestling team. Unfortunately, the cinematographer had never filmed wrestlers before, and he apparently didn't know much about physics either. He could not reliably determine when a wrestling hold would lead to one or the other wrestler's advantage. During the sparring sessions there was one move that I had started on top, but I planted my center of mass too high and my support points became controlled by my opponent. I executed a failed hold on his arms and torso that ended in my getting flipped to my back and pinned. Sure enough, this was the segment they edited into the program, and it was viewed by millions of people. When I asked the producer about it later he replied, "But you looked like you had him!"

Being a good, or even an average, wrestler ensured that I was in good physical condition for more than fifteen years of my adult life. This fact apparently did not go unnoticed by the public information office of Columbia University, who tracked me down at Princeton to recruit me for the 1997 Studmuffins of Science calendar. The invitation arrived via e-mail.

Date: Mon, 19 Feb 1996 11:33:07 -0500 (EST)
From: Robert J Nelson
To: ndt@astro.Princeton.EDU
Subject: Studly Scientists

Hi Neil!

Bob Nelson here at Columbia University's Office of Public Information.

You may have heard of the Science Studmuffins calendar. Karen Hopkin, a producer for Science Friday on NPR, did this calendar for 1996 and is back looking for more scientists to appear next year. Let me immediately add that everyone wears clothes in this particular calendar and I have one hanging in my office at Columbia. Although Ms. Hopkin emphasizes physical attractiveness, I think she's looking for well-rounded scientists (well, probably not in the physical sense!!) who have diverse interests and give the lie to nerd stereotypes.

Anyway, if you're interested, let me know, or maybe the PR office at Princeton will be interested.

Cordially,
Bob Nelson
Office of Public Information and Communications
Columbia University

My e-mail reply was brief:

From: Neil deGrasse Tyson <ndt@astro.Princeton.EDU>
Date: Fri, 8 Mar 1996 18:52:53 -0500
To: Robert J Nelson
Subject: Studly Scientists

Dear Bob,

Thank you for the flattering invitation to participate in the 1997 "Studmuffins of Science" calendar.

I have worked hard to be respected for my mind rather than my body. I believe I have finally succeeded, and thus do not wish to jeopardize my long-fought efforts.

Good luck, nonetheless, in your recruitment.

Sincerely
Neil deGrasse Tyson
Princeton Astrophysics

I remain flattered by the request but have no regrets for declining the invitation. If I wasn't going to dance half nude to "Great Balls of Fire" back in Texas, then I was not going to pose as a studly scientist for a nationally distributed calendar.

That being said, just four years later, in the summer of 2000, I was selected by *People* magazine to appear as the Sexiest Astrophysicist Alive for its annual Sexiest Man Alive double issue. And I agreed to it. This time, however, I was vastly more established as a scientist and an educator and I felt that my career would survive the designation. *People* magazine selects thirteen men for the Sexiest Man Alive article in the Sexiest Man Alive issue. One of them transcends category. He becomes *the* sexiest man alive and gets pictured on the cover. For my year, that designation went to Brad Pitt. For the rest of us, we represent specific categories, some repeating from year to year, some not. Repeating categories include sexiest action star, sexiest news anchor, sexiest author, and sexiest politician. Astrophysics is, you might have guessed, a nonrepeating category and might be swapped out from one year to the next with sexiest geologist, sexiest oceanographer, or sexiest accountant. I will not soon forget the interview, which was stuffed with questions like, "Who is your stylist?" "Which designer clothes do you like?" "Do women follow you around and try to make conversation?" Then she wanted the names and phone numbers of my wife and of old girlfriends and others who would speak to my sexuality. I was out of my league here, so I gave them the name and phone number of Sandi Kitt, the Hayden Planetarium librarian who happens to write romance novels on the side—more than twenty at last count. I figured Sandi could cook up quotable replies for whatever this interviewer wanted to know. What made it into print, however, was a quote from my wife. The interviewer had asked her reaction to when I first tried to take her up to the roof to see my telescope. My wife, a mathematical physicist, remained unimpressed with the invitation, and required me to deliver a cleverer line of courtship.

I still don't know where to put the Sexiest Astrophysicist Alive title on my curriculum vitae. But I do know the world contains relatively few astrophysicists. Among them, I knew I had Steven Hawking beat, but after that, the winner is not so obvious and was surely influenced by my overall visibility as director of New York City's Hayden Planetarium. So, once again, it's hard to get bigheaded about this sort of thing, but the designation continues to bring fun ribbing from colleagues and friends.

Knowledge and execution of the laws of physics can make you appear far more powerful than you actually are. While I was still in graduate school, during the days that followed an astrophysics conference on Italy's Amalfi Coast, my wife and I took a local bus tour of the many shops and restaurants of the nearby towns. As you might expect, the single road that connected all towns was narrow, with many switchback turns that barely negotiated the rocky coast. During one excursion our bus could not enter a town because a car had been sloppily parked, head first and askew, on the curb of our tight turn. Our bus honked its horn long and loud, but whoever owned the car was nowhere to be found. Traffic was building. Soon about twenty impatient cars collected behind us, wrapped beyond the previous turn. After ten minutes of futile steering gesticulations by townspeople to our bus driver, he finally gave up. He turned off the bus, sat in the opened doorway, and lit a cigarette. All the while I had been plotting a solution to this dilemma and had just received my cue. In this part of Italy, few people were within four inches of my height, fifty pounds of my weight, or within one hundred shades of my skin color. So I stood out just by being there. But now I would be remembered forever. I stood up and walked to the front of the bus. I exited and walked over to the problem car. Next, I bent down and deadlifted its rear section with a firm double grip on its bumper. I then slid the car sideways about three feet, providing adequate clearance for the bus to proceed. The forty or so people who were on the scene had stared silently at me during the episode. After I moved the car, they spontaneously burst into cheers and applause.

I thought nothing of the feat at the time, but reflecting upon it later I surmised that it might be the stuff of local legends. It had all the ingredients of a story that would pass from generation to generation. And it would not be immune to exaggeration. I can see it now. "The Legend of the Strong Man: A stranger from Ethiopia, who was as large as an ox, came to our town. He was the silent type. No one knew his name. He was a drifter. But just when he had arrived, the parking brakes of a local bus gave way on the hill. As the bus began to roll, little Guiseppe was crossing the street while holding hands with his grandmother. The Ethiopian stranger reacted quickly by thrusting his massive body in front of the bus, stopping it with his bare hands before he lifted its front end and swung it from its deadly path."

First, European-made cars in small-town coastal Italy are relatively light. Second, sloppily parked cars along Europe's narrow streets tend to be parked headfirst, and most of them have engines in the front, so the light end (the rear) is what sticks out to block traffic. A human can raise more weight with a dead lift than with practically any other unassisted method. In a dead lift, you invoke primarily the muscles in your thigh, the body's strongest. Neither your shoulders, nor your arms, nor your back actually do work against the force of gravity. The world record dead lift is about half a ton. Using a dead lift to raise the light end of a car does not compete with raising the light end of a wheelbarrow, but it comes close.

The friction between rubber and concrete is among the highest between any two surfaces, which means if you want to slide a car sideways on a road then you must apply nearly as much sideways force as the weight of the car itself (an essentially impossible task). The secret is to apply an upward force on the rear of the car until the weight on its tires is less than your own weight. You can then nudge the car, inch by inch, in whatever direction your traffic needs require. The car probably weighed fifteen hundred pounds (tops) with two-thirds of that weight sitting primarily over its front tires. By applying an upward force of two or three hundred pounds, sliding the car sideways became a trivial exercise in the laws of physics.

I have not gone back to see if the local townspeople erected any statues to commemorate the perceived feat of strength. But enough time has elapsed (more than a decade) for the story to have become legend, if it's destined to become a legend at all.

With the major network news headquarters located less than a mile from the Hayden Planetarium, I am an easy date for them to get a quick sound byte on the latest discoveries in the universe. In February 1996, the discovery of a new extra-solar planet was announced, and ABC News sent a crew to the Hayden to solicit my comments for the evening news with anchor Peter Jennings. My comments would be part of a larger story on the subject that included interviews with the discoverers themselves, and others. We do not observe extra-solar planets directly. We infer their presence from gravitational effects on the host star, which we observe as a wobble in the star's position in space. I offered the interviewer one of my best explanations for how you deduce the existence of a planet using the Doppler shift of the star's spectrum. I further commented that the star's wobble in reaction to the planet's gravity is more accurately described as a jiggle, and I reenacted what the star does using my hips.

In spite of what I thought was an erudite explanation of the discovery, perhaps expecting one or two sound bytes to be extracted from it, all the news showed of me that evening was my jiggling hips. For media interviews, I have since refrained from using my body to assist my explanations of scientific phenomena.

Just a few months later, in May 1996, I was coincidentally in Washington, DC, with other high-level administrators from the American Museum of Natural History to plumb for mutually beneficial projects with NASA in a meeting at their headquarters. Of the many museums across the land, NASA had no particular reason to presume that we were different or special. Our encounter with NASA officials was polite and cordial.

On our way out of the NASA headquarters, the cell phone of one of our team members rang. It was the communications department of the museum. They received an inquiry from ABC's *Nightline* about a major discovery on Mars, and inquired whether I would be interested in appearing that night to discuss it. I knew where all of NASA's space probes were in the solar system at that moment, and none of them were positioned to make a breakthrough discovery about Mars. After fifteen minutes of phone calls between us and the museum and between the museum and the *Nightline* producers, I gleaned that NASA was about to announce that extraterrestrial life may have been discovered on Mars. It could only have come from the analysis of meteorites. I agreed to do the interview, provided that I could be supplied with the original research paper and time enough to study the results. In almost every case where I am called upon to comment on one cosmic discovery or another, the major networks will have also interviewed the scientists responsible for the results. This frees me to offer big-picture perspectives on the role and meaning of the discoveries to the typical viewer.

No time remained to fly back to New York City that evening and be interviewed, so I was interviewed at the ABC affiliate studios in Washington, representing the Hayden Planetarium of the American Museum of Natural History. The segment producer handed to me what looked like a bootlegged galley of the original research paper, and I had about ninety minutes to study it. You should know that the paper was professionally researched, and was written with language that was as humble and tentative about the findings as you could imagine, but you would never know it from the headlines that were to come.

The next day, the president of the United States, in a move I had never before seen, introduced the NASA press conference from the lawn of the White House. The head of NASA gave an introduction of his own that included one or two scientific recollections from his childhood, including trips with his father to New York City's Hayden Planetarium. I don't know if he would have mentioned the

Hayden anyway or if he mentioned it only because we were in his face the day before, but it was a warm gesture felt by all New Yorkers who were watching.

The day of the NASA press conference, the CBS evening news interviewed the head of NASA, the lead author of the research paper, Carl Sagan (by telephone), and me. I was flattered and honored to be part of that threesome, but I was especially happy to offer comments and perspectives that might further enhance the scientific appreciation for the discovery in the hearts and minds of the millions of Americans who were watching. The media frenzy was, I believe, appropriate to the significance of the news story. Conspiracy skeptics were certain all the hoopla was an overnight stunt to reinvigorate NASA's diminished funding from Congress. NASA funding did receive a small uptick, but the naysayers clearly had not seen the original research paper, which had been years in the making.

I was not alone in my expertise. A half dozen scientists at the museum had knowledge that could be tapped for interviews about the Mars rock. We have biochemists, meteorite specialists, and solar system experts. And almost weekly, one scientist or another from the dozen research departments is consulted by the media about a breaking scientific story. The American Museum of Natural History is not just another museum.

On another media occasion when the networks were hungry for astrophysicists, the infamous asteroid 1997 XF11 was reported as having a real chance of striking Earth. Predictably, a media frenzy followed. The date was March 11, 1998. Within twenty-four hours, on March 12, the threat of impact was retracted after more detailed calculations became available—the asteroid would miss Earth by six hundred thousand miles. The producers for ABC's *World News Tonight* promptly called on me to explain what the hell had just happened. They wanted Peter Jennings, their ace anchor, to conduct a live, on-air interview, which for me was without precedent.

So I donned my best dark suit, my best French-cuffed shirt, and my favorite astro-novelty tie, and showed up at the ABC network studios on West Sixty-seventh Street in Manhattan. Of the fifty or so ties that I own, about half have patterns that evoke astronomical themes. Some are nerdy like the one with a space shuttle being launched straight up the tie. Others are artsy, like the mockup of Vincent van Gogh's *The Starry Night*, complete with the pointy and wavy bush, the church steeple from the town, a few of the fuzzy stars, and the crescent moon—all deftly rearranged to look as though van Gogh painted them on a vertical strip of cloth. I own yet another category of astro-tie that is simply loud. This was the category of tie I wore for the Peter Jennings interview. That particular tie displays randomly oriented golden yellow stars, moons, and comets floating on a satin-black background.

During the interview, which lasted a typical two and a half minutes, Jennings asked about how the asteroid was discovered, why it was initially perceived to be a threat, and why all was now okay. At the end of the interview, Jennings somehow felt compelled to comment. He uttered two simple words in front of his 2 million viewers: "Nice tie!"

Afterward, on my way out of the studios, no fewer than a dozen people from the production staff, including writers, editors, and camera operators, came up to me to get a closer look. They were all astonished that Jennings broke script, which they asserted he hardly ever does. Over the twenty-four hours that followed, I received dozens of e-mails from all kinds of people—strangers and friends—each offering congratulations on the interview and humorously ending with the identical compliment, "Nice tie!" One e-mail happened to come from my eleventh-grade English teacher, Mr. Bernard Kurtin, who was famous for his witty cynicism. Once, when I cut class on one of the numerous Jewish holidays in September, one of my classmates told me that when he took attendance and noticed my absence, he asked the class, "Where is Rabbi Tyson?" I hadn't seen or heard from Mr. Kurtin since high school.

Of the sixty e-mails I receive per day, I will never forget his one-line message the evening of my appearance with Peter Jennings:

Date: Fri, 13 Mar 1998 09:32:36 +0000
To: tyson@astro.amnh.org
From: "Bernard Kurtin"

I thought your necktie was just so, so.

—Bernard Kurtin

I own an asteroid. Rather, I own a piece of an iron-nickel asteroid that slammed into Earth at a speed of several miles per second, was collected by a meteorite hunter, and made its way to an auction house in New York City, where I put forth the winning bid. This particular asteroid fragment weighs a couple of pounds and spans the palm of my hand. Striations across its face betray an explosive episode somewhere in the asteroid's journey through 4.6 billion years. At a gavel price of $1,300, it qualifies as the most expensive paperweight I have ever owned.

At the same auction, a second, larger meteorite piqued my interest. It too had an iron-nickel composition but weighed about fifteen pounds and was the size and proportions of a discus. This particular meteorite had an esthetic quality to its shape—a natural hole had smoothly worn through the center so that when mounted upright, the meteor looked like a stylized doughnut that could easily pass for an objet d'art. Apparently, I wanted it more than anybody else in the room because I soon became the lone bidder against a person who was posting live bids to the auctioneer via telephone from California. The unidentified caller and I leapfrogged right up to my spending limit, and then some, but my pocketbook was evidently no match for the phone-bidder's interest level and resources. Several weeks later I learned from the auction house that my opposing bidder was a famous producer of science-fiction films. Clearly, no matter how much money I had planned to spend that afternoon, I was not going home with that meteorite. My disappointment eased, however, when I realized that at least one great filmmaker shared my interest in acquiring an extraterrestrial of verifiable provenance.

Auctions notwithstanding, in 2001 the International Astronomical Union named an asteroid in my honor, out of respect for my continued efforts to bring the universe down to Earth. David Levy, patron saint of comet and asteroid hunting, discovered this particular asteroid and proposed my name for it without my knowing. He did this even after we found ourselves on opposite sides of the debate to demote the planet Pluto from its standing as a bona fide planet among the others of the solar system. Levy had authored a biography of Clyde Tombaugh, who discovered Pluto. It is obvious then, which side of the debate Levy was on. My asteroid orbits in the asteroid belt, about two and a half times farther than Earth from the Sun, along with tens of thousands of others. It's officially called 13123 Tyson, but one can't get too bigheaded about this, given that 13,122 asteroids before mine got named after some other person, place, or thing. I have nonetheless enjoyed the distinction, and I'm glad, last I checked, it's not headed for Earth.

Of the cataloged asteroids in the solar system, several can be seen and tracked with a small telescope. The word "asteroid" translates to "starlike," because, apart from their incessant motion against the background stars, they look much like ordinary stars. Planets, on the other hand, are bright and clearly identifiable through a telescope as celestial orbs—entire worlds beyond our own.

For most of the year, the planets Venus, Mars, Jupiter, and Saturn shine brighter than nearly every star in the sky. This means they tend to be the first to "come out" after sunset. (You now have a plausible reason why your wishes generally don't come true whenever you wish upon a star in the early evening sky.) The planets visible to the unaided eye orbit the Sun in periods that range from as brief as 88 days for Mercury, through as long as 29.5 years for Saturn. From month to month and from year to year, different planets will masquerade as the first star of the evening. Since all the planets orbit close to the plane of the solar system, every now and then two or more of them come into alignment on the sky when viewed from Earth. By alignment, I mean within a few degrees of each other so that they fit nicely in the field of household binoculars. The alignment of planets is no more rare than the exact configuration of all planets at any arbitrarily selected moment. Planetary alignments just happen to be more beautiful. While I was a postdoctoral research scientist at Princeton University's Department of Astrophysical Sciences, I received a phone call from a graduate student in the Chinese Studies Department. He was translating an ancient manuscript that chronicled a cosmic event that led to the overthrow of a dynasty. The student suspected that the cosmic event was an alignment of the planets but he wanted verification. So I invited him by.

I own several planetarium-style sky programs that run on my office computers. They all show you the configuration of the Sun, Moon, and planets, for any place on Earth and for any time of day, day of the year, and year on the calendar for thousands of years into the past and future. Some programs are better than others and retain their accuracy over a longer base of time. The graduate student translated the Chinese calendar dates and declared that the auspicious cosmic event must have fallen somewhere between 1960 and 1950 BCE on the Gregorian calendar. To be safe, I conducted a search of planetary alignments over a broader range of years, from 2000 to 1900 BCE. Not knowing which planets would participate, nor what separation would be ominous to the ancient Chinese, I selected for any combination of Jupiter, Venus, and Mars and looked for a mutual separation of less than twenty degrees on the sky. I dropped my calculator when I discovered that during the early morning hours of February 25, 1952 BCE, the five planets visible to the unaided eye—Mercury, Venus, Mars, Jupiter, and Saturn—all fell within a three-degree circle of each other in the dawn sky. If you isolate the planets Mars, Mercury, and Venus, they fit within an even tighter half-degree circle. The three-degree separation is so small that if you held your hand at arm's length, your thumbnail would eclipse all five of them. If people needed an excuse to overthrow a dynasty, they had found it. In my continued searches, no other time from 3000 BCE to 3000 CE produced such an impressive conjunction of the naked-eye planets.

I was fresh off the discovery of the Chinese planetary alignment when it was time to replace the Hayden Planetarium's Zeiss model VI star projector. It had been installed in the late 1960s, and it was ready for an upgrade. A team of us combed the world for a modern projector to replace it. One such trip was back to Zeiss in its Jena, Germany, headquarters to see a prototype of the latest model VIII projector in Zeiss's planetarium test dome. In the technological counterpart to kicking a car's tires before you buy it, I asked the engineers to take me to 1952 BCE. This happened to fall outside their test algorithms, but they attempted it anyway. With the naked-eye planet projectors on full zoom and their whirling motion against the background stars in a countdown of years, we were all relieved when all planets found each other and huddled together tightly on the morning of February 25. Their engineers were pleased, as was I. I now invoke the Chinese alignment as a test of all software and star projectors that I come to evaluate.

I'm a fan of the planets in any combination. When I was born, Mercury, Venus, Jupiter, Saturn, Uranus, Neptune, Pluto, the Sun, and the Moon were all in the sky. The planets normally bring me good luck—even though I don't believe in luck. But the week before the presidential elections of 1996, the NBC *Nightly News* with Tom Brokaw ran a series of spots called "Fixing America," in which various well-known and not-so-well-known people were interviewed for their perspectives on what was wrong with America. The people were further prompted for ideas about how they might remedy the problems. When I was chosen for one of those spots, the camera crew elected to film me in the Sky Theater of the Hayden Planetarium. For the scene, the Zeiss star projector was to my left, and a large, zoomed video image of Saturn was projected on the dome and floating above my right shoulder. I was finally photographed together with my favorite planet. I gave my best advice for the nation that day, declaring that science literacy was good, and even necessary, for the electorate to make informed decisions—about issues in modern society—that affect our lives. I went on to declare that public interest in cosmic discovery is high and should serve as a magnet for children's interest in science. When accompanied by Saturn, the subject of my desk lamp, how could anything go wrong with the interview?

These daily "Fixing America" segments typically featured two or three people. Out of curiosity, I asked the producer of the segment who else they were interviewing. I would be grouped with a football coach from a midwestern college, and the Metropolitan Opera singer Jessye Norman.

The following day, the segment aired on the news in the sequence: Jessye Norman, me, the football coach. The footage of me in the theater, flanked by Saturn and the star projector, looked otherworldly—as though I were visiting Saturn in its orbit rather than an image that Saturn was visiting me on Earth. But none of that mattered. Everything that I said was totally eclipsed by Jessye Norman. Her message that day was in direct response to the rhetoric of the Republican presidential hopefuls, who, during that election season, kept referring to how the poor needed to pick themselves up by their bootstraps. Bootstraps became the metaphor for programs that would reduce the welfare rolls.

Jessye Norman's visage radiates bright cheeks, a high forehead, and expressive, high-arched eyebrows. Her elocution reminds you of a classical orator. Her dignity, stately manner, and her gown could grace any throne in the world. Her intelligence and clarity of thought was manifest.

Ms. Norman enunciated all her syllables with poetic drama that only an opera singer can deliver: "We should dare to care about one another. We should not allow politicians to suggest that the poor are not our concern. These are the people who are not able to pull themselves up by bootstraps—they are not wearing any boots."

Cut. Print. End segment. End newscast.

The moment she spoke, I knew that my pleas for science literacy would pale by comparison. They should have left me and my Saturn for another day, or another occasion, or on the editing room floor.

Another planet that has figured prominently in my life, especially lately, is Pluto. I published an essay for *Natural History* magazine in February 1999 titled "Pluto's Honor." The time was not arbitrary. That month, Pluto regained its status as the most far-out planet after a twenty-year stint orbiting closer to the Sun than Neptune. Pluto's uniquely elongated orbit happens to cross Neptune's orbit for twenty years out of a 248-year period. In my essay I presented the case for demoting Pluto from its long-held status as a planet to a classification that aligns it more closely with comets found in the outer solar system. The argument is simple. Pluto never really fit into the family of planets. It is the smallest among them—indeed seven moons of other planets, including Earth's moon, are bigger. More than half of Pluto's volume is ice, so that if you brought Pluto closer to the Sun, say, the Earth-Sun distance, then Pluto would grow a hundred-million-mile-long cometary tail. Now what kind of behavior is that for a planet? The nails in the coffin come from the 1992 discovery of icy bodies beyond the orbit of Neptune that have more properties in common with Pluto (orbit, composition, size, etc.) than either Pluto or these icy bodies have in common with any other planet. We are left with little choice but to give Pluto its walking papers and require that it join this other class of objects. But all is not lost. Pluto would go from being the tiniest planet to being the largest known icy object in the outer solar system.

By about February 10, the mail started rolling in. I knew Pluto was popular among elementary schoolkids, but I had no idea they would mobilize into a "Save Pluto" campaign. I now have a drawer full of hate letters from hundreds of elementary schoolchildren (with supportive cover letters from their science teachers) pleading with me to reverse my stance on Pluto. The file includes a photograph of the entire third grade of a school posing on their front steps and holding up a banner proclaiming, "Dr. Tyson—Pluto is a Planet!" One of the letters was from the Pluto Protection Society, based near the Lowell Observatory in Arizona, home of the original photographic search that led to Pluto's discovery in 1930. And a newspaper article and profile on me, published the following month in the *New York Observer,* led with a small front-page headshot (indicating a larger article within), and the tagline "The Man Who Would Demote Pluto."

The world's arbiter of astronomical nomenclature and classification systems is the International Astronomical Union (IAU). It assembles committees that express learned scientific points of view that occasionally blend with political will. To offend the fewest people (unlike what became of my *Natural History* essay), the IAU straddled the fence on the issue, allowing people to call Pluto a planet while simultaneously accepting the growing (and irreversible) movement to classify Pluto as a comet. In the meantime, I will continue to x-ray packages sent to my office from third graders.

For most of recorded history, Earth was not thought to be a planet. Officially, planets were all those things in the sky—there were seven of them, including the Sun and Moon. Earth was a unique, stationary object, around which everything in the universe turned. The early pagan civilizations of Mesopotamia believed that all objects in the heavens were gods, but the most powerful gods were the seven planets. These supergods ranked by their speed across the sky—the slower they moved, the more ancient and powerful they were. With Saturn the slowest and the Moon the fastest, the seven planets were ranked as follows: Saturn, Jupiter, Mars, the Sun,

Venus, Mercury, and the Moon. The Mesopotamians, and later the Romans, assigned each planet to rule Earth's affairs, in sequence, hour by hour, for every day. Whichever planet happened to rule the first hour of a twenty-four hour day was that day's reigning planet.

An easy way to reconstruct the days of the week, in their familiar sequence, is to lay down a circle of the planets in sky-speed order:

Saturn

Moon   Jupiter

Mercury   Mars

Venus   Sun

Start anywhere around the circle (using the first spot as the first day) and count clockwise twenty-three more planets, one for each hour in the "current" day. The twenty-fifth planet rules the next day. Restart your counting from one, and continue this numerical ritual seven times in a row. You will recover all seven days of the week in calendar order: Saturn-day, Sun-day, Moon-day, Mars-day, Mercury-day, Jupiter-day, and Venus-day. When you substitute "Sabbath" for Saturday and "Lord's day" for Sunday, you get the basic Latin forms of the French, Italian, Spanish, and Portuguese name for the days of the week: Sabbata, Domenica, Luna, Martis, Mercurius, Jovis, Veneris. For some Western languages (English included) substitute the Anglo-Saxon gods Tiu, Woden, Thor, and Frigga for their Roman counterparts to get Tuesday, Wednesday, Thursday, and Friday.

The prevalence of religious mythologies among scientifically ignorant cultures that flourished millennia ago makes it easy to see why they would believe that planet gods exerted divine influence on human affairs. It's much harder to see why similar beliefs persist today, and every day, in the astrology pages of the newspapers, unless we are to admit to ourselves that contemporary society remains widely uninformed in matters of science. We fail in even the simplest of all scientific observations—nobody looks up anymore. Why else would people be surprised to learn that the Moon also comes out in the daytime; that the North Star is not, and was never in contention for, being the brightest star in the nighttime sky; that for most of Earth's population the Sun has never appeared directly overhead at any time of day or on any day of the year; that most of the eighty-eight constellations in the sky are wholly unrecognizable patterns when compared with the creatures and objects that legends and mythologies declare them to be; and that the planets journey back and forth across the sky, from one side of the Sun to the next, getting brighter and dimmer and brighter again.

We live in the days of evening distractions that include television, multiplex cinemas, and even books that can be read by electric light after dark. When I was in graduate school at Columbia University, an elderly woman with a strong Brooklyn accent called my office to ask about a bright glowing object she saw "hovering" outside her window the night before. I knew that a few planets were bright and well placed for viewing in the early-evening sky, but I asked more questions to verify my suspicions. After sifting through answers like, "It's a little bit higher than the roof of Marty's Deli," I concluded that the brightness, compass direction, elevation above the horizon, and time of observation were consistent with her having seen the planet Venus. Realizing that she has probably lived in Brooklyn most of her life, I asked her why she called then and not at any of the hundreds of other times that Venus was bright over the western horizon. She replied, "I've never noticed it before." You must understand that to an astrophysicist, this is an astonishing statement. I asked how long she has lived in her apartment. "Thirty years." I asked her whether she has ever looked out her window before. "I used to always keep my curtains closed, but now I keep them open." Naturally, I then asked her why she now keeps her curtains open. "There used to be a tall apartment building outside my window but they tore it down. Now I can see the sky and it's beautiful."

I have similar encounters with all sorts of people about once per month. When it's not Venus, it's Jupiter or Mars. And when it's not planets, it's odd cloud formations or bright shooting stars. With phone calls such as these, with people taking the time and energy to ask about what they do not understand, I have a renewed hope that society can shed its superstitions and embrace the enlightenment that comes from just a basic understanding of how the universe works.

The supply of professional astrophysicists in the world has held for some time at a steady ratio of one in a million people, which is not nearly enough. But it does mean that if you find yourself sitting next to an astrophysicist on the airplane, then you had better ask all your pent-up questions about the universe. You do not know when your next encounter will be.

In addition to the current total of astrophysicists, we clearly need one astrophysicist for every disaster movie produced, and then some. With the nation's urban murder rates falling to half-century lows, the motion picture industry can no longer depict crime as a stereotype of life in the city. But unlike romantic comedies or action-adventure thrillers, most disaster films tap scientific arteries of knowledge for their storylines. Deadly viruses, out-of-control DNA, evil aliens, monsters, and killer meteors are recurring themes in apocalyptic films. Unfortunately, a film's scientific literacy hardly ever measures up to its plot, leading to unforgivable abuses of the way the world works.

I'm not talking about simple bloopers such as when a Roman centurion sports a wristwatch while riding a chariot. Or when the shadow of a microphone boom creeps into a scene. These mistakes are inadvertent. I'm talking about purposeful yet ignorant bloopers, like reversing the sunset to pretend you have filmed a sunrise. Are cinematographers too lazy to wake up before sunset and get the real footage? Sunrise and sunset are not time-symmetric events.

Or how about when they show Christopher Columbus on the deck of the *Santa Maria*, peering through a telescope that was not invented for another 116 years?

And why did James Cameron, the talented director of the 1997 film *Titanic*, take the time to get every imaginable detail correct— from the number of rivets in the hull, to the patterns in the dinner plates—yet he got the wrong nighttime sky? What might have been the constellation Corona Borealis (the Northern Crown) is shown overhead on that fateful night. But it has the wrong number of stars. Why? Had Cameron attended Camp Uraniborg this mistake might not have happened. I'd bet he researched the costumes to be pre-cisely the styles of the period. If someone had been onboard wearing love beads, bell-bottom jeans, and a large Afro, you know that viewers would have complained loudly that Cameron had not done his homework. Am I any less justified in my outcries?

My gripes are not just with Hollywood. What about those majestic stars in the ceiling of New York City's Grand Central Ter-minal? It's a canopy of constellations rising high above the heads of hurrying commuters who haven't got time to look up anyway. But the star patterns are backward when compared with the real night sky. Rather than just admitting the mistake, a sign in the lobby tells us, "Said to be backwards, [the ceiling is] actually seen from a point of view outside our solar system." But a second error has now been committed in an attempt to cover up the first: no point of view in our galaxy will reverse the constellation patterns of Earth's night sky. As you leave the solar system, and travel among the stars, all that happens to Earth's constellations is that they become scram-bled and wholly unrecognizable.

What society needs are scientifically literate reviewers. Why should a theater critic be limited to making critiques such as, "the characters stretched credulity," or "the tonal elements clashed with the emotional flavor of the set designs"? Just once I want to hear a critic say, "The Scarecrow botched his recital of the Pythagorean theorem when the Wizard gave him a diploma" when reviewing the 1939 fantasy classic *The Wizard of Oz*. A critic might also declare, "Flying saucers traveling quadrillions of miles through interstellar space don't need runway lights to land on Earth" when reviewing the 1978 almond-eyed alien classic *Close Encounters of the Third Kind*. I would have loved for a critic to notice that the Moon phases grew in the wrong direction throughout Steve Martin's otherwise-charming 1986 romantic comedy *LA Story*. And I would have rejoiced had I heard just one critic say, "A killer asteroid the size of Texas would have been discovered two hundred years ago, not two weeks before impact" when reviewing the 1998 summer block-buster *Armageddon*.

Only when such errors are highlighted will the public begin to appreciate the inescapable role that the laws of physics play in everyday life.

If you want to write a book, make a film, or engage in a public art project, and if this work makes reference to the natural world, just call your neighborhood scientist and chat about it. When you seek "scientific license" to distort the laws of nature, or when you want to corrupt the appearance of night sky phenomena, then I prefer you did so knowing the truth, rather than inventing a story-line cloaked in ignorance. You may be surprised to learn that valid science can make fertile additions to your storytelling—whether or not your artistic objective is to destroy the world.

Practically every scientific claim ever made was, or should have been, accompanied by a tandem measure of the reliability of the claim. When reporting scientific discoveries, the popular press hardly ever conveys these inherent uncertainties in the data or the interpretation. This seemingly innocent omission carries a subtle, misguided message: if it's a scientific study, the results are exact and correct. These same news reports often declare that scientists, having previously thought one thing, are now forced to think something else; or they are forced to return to the mythic "drawing board" in a stupor. As a consequence, if you get all your science from press accounts then you might be led to believe that scientists arrogantly, yet aimlessly, bounce back and forth between one perceived truth and another without ever contributing to a base of objective knowledge.

But let's take a closer look.

New ideas put forth by well-trained research scientists will be wrong most of the time because the frontier of discovery is, for the most part, a messy place. But we know this and are further trained to quantify this level of ignorance with an estimate of the claim's uncertainty. The famous "plus-or-minus" sign is the most widely recognized example. We typically present a tentative result based on a shaky interpretation of poor data. Six months later, different, yet equally bad data become available from somebody else's experiment and a different interpretation emerges. During this phase, which may drag on for years or even decades, news stories implying unassailable fact get written anyway.

Eventually, excellent data become available and a consensus emerges—a long-term process that does not lend itself to late-breaking news reports. Studies on environmental health risks, or the effects of food consumption on diseases and longevity, are especially susceptible to being overinterpreted. The financial consequences of premature news stories, and the attendant reactions on Wall Street, can be staggering. In 1992 a Florida man brought a lawsuit against two cellular phone manufacturers by claiming that his wife's death from brain cancer was caused by her heavy use of cellular phones. When this and several similar claims hit the news in late January 1993, the market capitalization of publicly traded cellular phone companies fell by billions of dollars in less than a week. Several astrophysics colleagues were financially poised to take advantage of this lemminglike market reaction to the perceived hazards of cell phones. A little bit of analysis goes far: you can get brain cancer without ever using a cellular phone. And since the popularity of cellular phones was on the rise, you expect some users to die from brain cancer just as some users would die from heart disease, or from old age. In this case, there was no definitive study to establish a cause and effect between cellular phone use and brain cancer, yet people overreacted anyway. Fortunately, most of the comings and goings of astrophysics have so little impact on how people conduct their daily lives that I can spend more time joking about the problem than crying about it.

Initial uncertainty is a natural element of the scientific method, yet the scientific method is, without question, the most powerful and successful path ever devised to understand the physical world. When a published scientific finding is confirmed and reconfirmed and re-reconfirmed and re-re-reconfirmed, then further confirmation becomes less interesting than working on another problem. At that time, and only at that time, the new nuggets of knowledge are justifiably presented with little or no uncertainty in the basic textbooks of the day. Consistency and repeatability are the hallmarks of a genuine scientific finding. For if the laws of physics and chemistry were different from lab to lab, and from one moment to the next, then scientists would all just pack up and go home.

Occasionally, scientists ignore their uncertainties because, for the most part, scientists are people too. There are arrogant ones, lovable ones, loud ones, soft-spoken ones, and boneheaded ones. Every scientist, myself included, has colleagues who fill each category. In published research papers, however, we are typically timid because of the semipermanence of the printed word and because of the overwhelming frequency of wrong ideas. Most results flow from the edge of our understanding and are therefore subject to large uncertainties.

More often than not, a scientist's printed word presents an

honest, almost humble uncertainty that goes unnoticed when people reflect on the history of scientific misconceptions. What about that 1996 research paper that claimed to have found life in a Martian meteorite? Writing in the journal *Science*, the nine coauthors noted, among other things, in their abstract:

> The carbonate globules [in the Martian meteorite] are similar in texture and size to some terrestrial bacterially induced carbonate precipitates. Although inorganic formation is possible, formation of the globules by biogenic processes could explain many of the observed features . . . and could thus be fossil remains of past Martian biota.

From the oversized newspaper headlines that followed and the persistent media request for my time, you would never guess that the original research paper contained such unassertive language.

A rare, but now-famous case of a misreported uncertainty coupled with an overconfident claim by a scientist took place in early 1998, when the Central Bureau for Astronomical Telegrams (the clearinghouse for astronomers of the world who need to disseminate up-to-the-minute sky phenomena among colleagues) announced the discovery of a mile-wide asteroid whose orbit would bring it dangerously close to Earth in the year 2028. (Formerly sent around the world via telegram, these notices are now distributed instantly via e-mail.) The offending asteroid was coded 1997 XF11, which cryptically identifies when in the year 1997 the asteroid was discovered. This was the asteroid that prompted the infamous "Nice tie" remark from Peter Jennings on ABC's *World News Tonight*. The telegram reported on March 11, 1998:

> This object, discovered by J. V. Scotti in the course of the Spacewatch program at the University of Arizona on 1997 Dec. 6 . . . recognized as one of the 108 "potentially hazardous asteroids", has been under observation through 1998 Mar. 4. . . . An orbit computation from the 88-day arc . . . indicates that the object will

pass only 0.00031 AU from the earth on 2028 Oct. 26.73 UT! Error estimates suggest that passage within 0.002 AU is virtually certain, this figure being decidedly smaller than has been reliably predicted for generally fainter potentially hazardous asteroids in the foreseeable future.

When converted to everyday language, the announcement declared that the asteroid's most likely path would bring it within thirty thousand miles of Earth (a cosmic hair's width), but the uncertainty in the calculation allowed the asteroid to come anywhere within a two hundred thousand mile "error circle" surrounding Earth.

I remember reading this telegram from my office at Princeton University within hours after its release. All I could think to myself was: It was bound to happen some time. In the year 2028 I will be seventy years old. What a way to go! But then I was recoiled from an unfamiliar combination of emotions: one of shock, that life as we know it could end in my natural lifetime, and one of perverse pride in knowing the laws of physics that enabled us to make the prediction.

When the substance of the telegram was further distributed via press release from the American Astronomical Society, passing along the hair-raising words "virtually certain," a media deluge followed.

The telegram went on to give the best available coordinates for the object—obtained from observers who were tracking it—preceded by a scientifically sensible appeal: "The following ephemeris is given in the hope that further observations will allow refinement of the 2028 miss distance." The next day, on March 12, 1998, another telegram appeared that announced the existence of what astronomers call a "prediscovery" photograph of the asteroid, obtained from archival survey images taken in 1990. This significantly extended the baseline of observations to well beyond the original eighty-eight days. (Longer baselines always provide more accurate estimates than shorter ones.) Calculations that incorporated the new data narrowed the error circle to a skinny ellipse that handily shifted Earth from within the range of collision uncertainty to well outside of it. Five weeks later, a telegram was issued that corrected the alarmist language of the first announcement and admitted that the original telegram's uncertainties could have been sharpened if a more sophisticated method of calculation been used.

The episode was widely reported as a blunder, but at worst, the original calculation was simply incomplete. At best, it was a valid scientific starting point. True, the survival of the human species was involved, but most important, everything worked the way it was supposed to. The early estimate, and the better estimates that followed (within a day!) were a model of the scientific method and how it has the power to refine itself as our knowledge approaches an objective reality.

After what had been twenty-four hours of sensationalist journalism across the country, the retraction spawned sighs of relief. In particular, the *New York Post*, a colorfully written daily newspaper in New York City, ran the inimitable headline: KISS YOUR ASTEROID GOODBYE. And a few days later, an illustration by cartoonist Jesse Gordon on the op-ed page of the *New York Times* depicted the asteroid changing its collision course over a sequence of panels. We are treated to the top nine reasons why the asteroid has decided to not hit Earth, one of them being: "No desire to spend the rest of its days in the lobby of the Museum of Natural History."

How certain can we be of a scientific measurement? Confirmation matters. Only rarely is the importance of this fact captured in the media or the movies. The 1996 film *Contact*, based on the 1983 novel of the same name by the celebrated astronomer Carl Sagan, was an exception. It portrayed what might happen—scientifically, socially, and politically—if one day we make radio-wave contact with extraterrestrial intelligence. When a radio signal from the star

Vega rises above the din of cosmic noise, Jodie Foster (who plays an astrophysicist) alerts observers in Australia, who could observe the signal long after the stars in that region of the sky had set for Americans. Only when the Australians confirm her measurements does she go public with the discovery. Her original signal could have been a systematic glitch in the telescope's electronics. It could have been a local prankster beaming signals into the telescope from across the street. It could have been a local collective delusion. Her confidence was boosted only when somebody else on another telescope with different electronics driving an independent computer system got the same results.

The accuracy and integrity of the above scene almost makes up for a grave mathematical blunder earlier in the film. In a scene where Jodie Foster and her handsome love interest, Matthew McConaughey, take their first kiss, Jodie Foster recites the following line:

> If there are 400 billion stars in our galaxy, and only one in a million of them had planets, and only one in a million of those stars with planets had life, and only one in million of the stars with planets that have life, have intelligent life, that still leaves millions of planets to explore.

If you do the arithmetic, correctly, you are left with not "millions" of planets but 0.0000004 planets to explore. For this blunder, I don't blame the writers or producers. They have enough on their minds.

I blame Jodie Foster.

She must have rehearsed her lines many times over. There must have been multiple takes of the same scene, as is common in high-budget films. At some point she might have caught the error. Last I checked, she was a graduate of Yale. I'm pretty sure they teach arithmetic there.

I was one of the lucky few to attend the world premiere of *Con-*

tact in Pasadena, California, by the invitation of Anne Druyan, Sagan's widow and coauthor of the film's story—my first and only Hollywood world premiere. In attendance was also the astrophysicist Frank Drake, whose famous Drake equation was the subject of Jodie Foster's numerical recitation. During the infamous scene, Frank Drake did not convulse and was remarkably forgiving in his attitude and behavior, so I could not justify taking any more action than he did, which was nothing.

If Americans had more training with the metric system, and powers of ten in general, then the relationship between, and among, numerical quantities would become a matter of understanding rather than of rote memory. Compare the following sentence pairs: A kilometer is a thousand meters, a mile is 5,280 feet. A meter is a hundred centimeters, a yard is thirty-six inches. A liter contains ten deciliters, a quart contains thirty-two ounces. With powers of ten built into the metric system of measurement, had Jodie Foster been comfortable with it, she might have calculated what was going on in her lines rather than simply memorized them.

The system of royal governance was not the only thing overthrown during the French Revolution of 1789. So too was the system of weights and measures that was based on the length of various human body parts and on arcane references to nonstandardized containers. What replaced it was the decimal system, inspired by the fact that the average person has ten fingers and ten toes, which led to a counting system that uses ten numerals. If our species naturally had some other number of fingers we surely would be using a different counting system. I have no doubt that octopuses and arachnids do their arithmetic in base eight.

Today, anybody who needs to know the metric system knows it. This includes scientists, engineers, and international industrial corporations. Last I checked, only four countries are left in the world that do not officially sanction the metric system in their general population: Liberia, Myanmar, South Yemen, and the United States of America. America, the largest military, economic, and industrial force the world has ever seen, commonly uses inches, feet, and miles, and pints, quarts, and gallons for its daily measurements. A government initiative to convert Americans over to the metric system, begun in the 1970s, has largely failed. It's the sort of thing you need to accomplish cold turkey. Maps, road signs, and even baseball parks all posted dual distances that included the metric system, which meant nobody had to learn a thing.

I don't think that the Fahrenheit temperature scale will ever lose ground to Celsius in America. As arcane and logic defying as the Fahrenheit scale is, its ten-degree increments are too useful for weather forecasters to abandon: Simple proclamations such as "It'll be in the 60s today" or "The temperature will drop down into the teens tonight" or "The temperature will stay in the 90s today" have served to group temperature intervals into comfort levels. But other aspects of the scale, such as its low and unfamiliar zero point, have left some Americans hopelessly confused. A few years ago, I happened to be listening to the radio in midwinter, when the outdoor Fahrenheit temperature was in the low single digits. In New York City, temperatures that low are sufficiently rare to warrant fresh observations on the meaning and significance of air that cold. The temperature dropped slowly through the night, degree by degree. As the readings neared zero, the announcer declared with what was surely a straight face, "There is almost no temperature left."

In spite of isolated embarrassing moments such as those, I am happy to support our colloquial vocabulary, borne of the British system of units: Motion picture cameras will not stop shooting *footage* of film. *Mile*stones in people's lives won't all of a sudden go away. Running backs in American football will continue to gain (or lose) *yardage* on the gridiron. Car owners will never abandon the *mileage* measure of an engine's efficiency. We will surely not give up the use of *inch* as a verb. *Inch*worms should be environmentally protected from going metric. Small people will continue to be pint-sized. And an ounce of prevention will forever provide a pound of cure.

Unlike most of my colleagues, I am neither upset nor concerned that America is not entirely metric. One of my reasons is that America is largely metric already, although people don't realize it. Our money has one hundred pennies to the dollar. Our standard photographic film is 35 millimeters. The size of camera lenses, and filters that screw into them, are all measured in millimeters, as is the size of binoculars. All wine bottles and most liquor bottles are 750 milliliters and multiples (or fractions) of it. Piston displacement for car engines is now routinely measured in liters. One- and three-liter plastic bottles of soft drinks are industry standards. Nine-millimeter pistols are becoming the handgun of choice for urban law enforcement. The brightness of household light bulbs and the power of hair dryers are both measured in watts, a metric unit of energy consumption. The power of a car battery is measured in volts and amps, each a metric unit of electricity. Nearly all races run by athletes are measured in round-number meter-length distances. All prescription and over-the-counter medicines have the strength of their active ingredients measured in grams. And kilos are, of course, a standard unit of drug trafficking.

My primary reason for not caring whether America goes metric is that the British system of feet and Fahrenheit, acres and inches, and pints and pounds, carries some charming history, even if people who use the terms do not know it. When you are a visitor to America, our units of measure become one of the things telling you that you have landed in a country different from your own—a country with different history and different customs.

As a visitor to America, the absence of the metric system will, of course, be the least of your adjustment problems. The hodge-podge collection of words that comprise American English, along with paper money that is all the same size and color, might be a greater concern than whether you can interpret the Fahrenheit temperature on a flashing bank thermometer. When you visit another country you expect things to be different. That's the most common reason why people go on vacation—to be someplace different.

As long as baseball fields are measured in yards, hotdogs come in packs of eight to the pound, and the ingredients to grandma's apple pie are measured in cups and teaspoons, you know you are in America—land of the free and of the metrically challenged.

My mother tells me I was a relatively well-behaved child. Whether or not she has selective memory, this picture of me as a four-year-old might argue in her favor. (*Family Archives 1962*)

This album was given to me by my grandmother Altima deGrasse Tyson so I would collect and preserve memorabilia from each year in school. The report cards and other artifacts of evaluation paint a consistent picture: the absence of support from teachers who judged that I was not among their best students. (*Family Archives 2004*)

Teacher Comments

Period 1

Period 2 *Neil should cultivate a more serious attitude toward his school work.*

Period 3

Parent Comments

Period 1

Parent's Signature — *Cyril D. Tyson*

Period 2

Parent's Signature — *Cyril D. Tyson*

New Grade and Room — 4 – 1   2 / 5

First Day of New Term — 9/11/67

This report card is from the third grade at P.S. 81 in the Bronx. Confusing social energy with not being serious about school, Mrs. O'Connell complained to my parents about my approach to schoolwork. Comments such as these were typical of what teachers and later, professors, would say of me, all the way through graduate school. *(Family Archives 2004)*

During a district track meet in Van Cortlandt Park, at age twelve, the author poses (front row, second from right) with other medalists after winning silver in the 100-yard dash. This occasion was the last time I valued athletics over intellectual pursuits, having thereafter realized that I trained to be good at sports more because society expected it (as society did of all black children) than for any reason inherent within me. (*Family Archives 1970*)

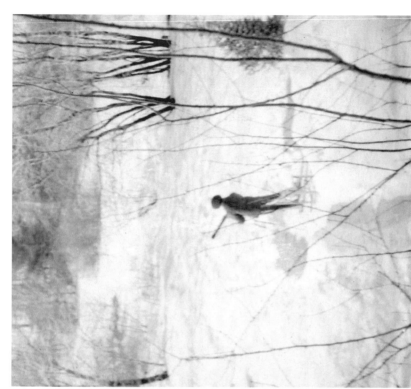

After a late-season heavy snowfall, I carved a path into our backyard in Lexington, Massachusetts, and I used my first telescope to track sunspots while measuring the Sun's rate of rotation. (*Family Archives 1971*)

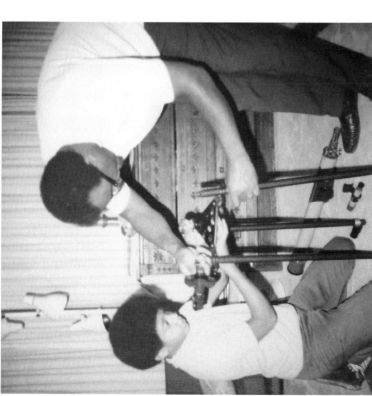

My father, Cyril deGrasse Tyson, assists me as I assemble my first telescope for my twelfth birthday. That year, while in seventh grade at Muzzey Junior High School in Lexington, Massachusetts—without a television and without the city streets calling me—I earned higher grades than at any time before or since. I finished the year with the highest GPA in the school. (*Family Archives 1970*)

I took my first astronomy course at the Hayden Planetarium at age eleven, and I continued taking courses through age fifteen. For each offering, the institution issued a certificate of completion. My first intellectual role model and mentor, Dr. Mark Chartrand III, taught my favorite course, Astronomy Roundtable, the certificate for which is pictured here. He was also head of the planetarium at the time. Now I, as director of the Hayden Planetarium, issue these same certificates, signing each one with an overpriced fountain pen, vowing to be as influential for others in my job as Dr. Chartrand was for me. (*Family Archives 2004*)

THE AMERICAN MUSEUM
—
HAYDEN PLANETARIUM

NEW YORK, NEW YORK

NEIL D. TYSON

*having been in regular attendance, and having
demonstrated satisfactory interest in and understanding
of the subject matter, is hereby awarded this*

CERTIFICATE OF COMPLETION

in

ASTRONOMY ROUNDTABLE

Chairman _____ , Instructor _____ Date 4 June 1974

The fifth largest ocean liner in the world at the time, the SS *Canberra* carried fifteen hundred scientists and their families on an expedition to the northwest coast of Africa to study the June 30, 1973, total solar eclipse. Lasting seven minutes and fourteen seconds, this eclipse was one of the longest on record. Five members of the scientific staff of the Hayden Planetarium were in attendance as instructors and lecturers. Among the many luminaries on board were *Apollo 11* astronaut Neil Armstrong and prolific science writer Isaac Asimov. *(Family Archives 1973)*

This photograph, taken two hundred miles off the coast of Senegal, shows a thicket of tripods on the eclipse day. At age fourteen, I was the youngest person on board without a guardian. But I was not without my telescope. Many years later, the *Canberra* would be used to ferry British soldiers to the southern hemisphere to wage war in the Falkland Islands. *(Family Archives 1973)*

At age fifteen, I was a student camper at Camp Uraniborg in the Mojave Desert, Southern California, a place that catered to kids who were astronomy nerds. We lived nocturnally, under cloudless skies, with a battery of telescopes at our disposal. Here, I am posing with binoculars next to a large-format astrocamera. *(Family Archives 1974)*

This view is from the roof of the Skyview apartments in Riverdale, New York, looking west after sunset, across the Hudson River and toward the New Jersey Palisades. My first view of the Moon and stars through binoculars took place on this vista at age nine. *(Family Archives 1974)*

To buy my second, much larger telescope, I earned money by walking dogs belonging to residents in the apartment complex in which I lived. Pictured are two—Tuffy and Ceba— of the six who were my regulars. At fifty cents per walk per dog, the money added up quickly. *(Family Archives 1975)*

Although the graduate school does not recognize a valedictorian, I was selected to offer reflective and parting remarks to my fellow doctoral candidates during convocation at Columbia's School of Arts and Sciences. During this speech I alerted the audience that my PhD in astrophysics that afternoon would bring the world total of black astrophysicists to seven. *(Family Archives 1991)*

On the subjects of cosmology and life in the universe, I was one of four panelists to join host Robert Kuhn in an hour-long PBS debate and discussion titled *Closer to Truth*. Television appearances such as this one have greatly magnified my daily efforts to bring the universe to the public. Pictured clockwise from the left: Caltech astrophysicist Roger Blandford, the author, Kuhn, Caltech provost Steve Koonin, and MIT cosmologist Alan Guth. (© 2000 *CLOSER TO TRUTH: Science, Meaning and the Future*)

This book signing at the American Museum of Natural History was for my fifth book on the universe, a collaborative effort with two colleagues (Charles Liu and Robert Irion), capturing my lifelong quest to connect the laws of physics that describe the universe to activities and events of everyday life. The book's title, *One Universe: At Home in the Cosmos*, and its role as the companion book to the February 2000 opening of the Rose Center for Earth and Space (containing the rebuilt Hayden Planetarium) reflect this quest. *(Jason Green Photography 2000)*

First Lady Hillary Rodham Clinton invited me and Dr. Marcia McNutt, president and CEO of the Monterey Bay Aquarium, to give tandem talks at the White House to a hundred or so dignitaries, heads of agencies, and guests as part of her Millennium Council lecture series. Our broader task that morning was to compare and contrast our professions. But above all, we celebrated the enterprise of exploration—from the ocean depths to the outer reaches of space. (*Official White House Photo 2000*)

Pres. George W. Bush's Aerospace Commission, of which I was one of twelve members, conducted research around the country and around the world to assess the health and climate of the American aerospace industry. During a visit to the Kennedy Space Center, we spent several hours comparing notes in the launch-site bunker of the Mercury program. Commissioners are seated from left to right at the table: Ed Bolen, president and CEO of the General Aviation Manufacturers Association; Whit Peters, former secretary of the air force; Bob Walker, commission chair; *Apollo 11* astronaut Buzz Aldrin; and the author. Various staffers from supporting government agencies (Dept. of Energy, Dept. of Defense, NASA, Dept. of Commerce, Dept. of Education) are seated in the rear. *(Family Archives 2002)*

As commissioner, you're allowed to lean on the missiles. As part of our world tour of the international aerospace industry, one of our stops was the Farnborough Air Show, which alternates each year with the more famous Paris Air Show. On display is a Navy F/A-18 Super Hornet fighter jet, carrying an assortment of weapons under its wing. The one under my arm is a High-speed Anti-Radar Missile (HARM). The long one protruding behind me is a laser-guided Paveway bomb. *(Family Archives 2002)*

On this commission trip to Beijing, my first visit to China, I was quite impressed with a perfect cellphone connection to my parents in Westchester, New York, placed from the Great Wall of China. Our final report contained information about China's extraordinary rate of growth in aerospace technology. However, few paid attention, preferring instead to think of China as a country of bicycles and peasants. Eighteen months later, China became the third country (after the United States and Russia) to launch a man into orbit. People are listening now. *(Family Archives 2002)*

On the commission trip to Moscow, my first visit to Russia, we toured Star City, the Russian equivalent of the Kennedy Space Center. Posing for the photo in the cramped *Soyuz* capsule, from left to right, are Bob Walker, the author, and Robert Stevens, then president of the Lockheed Martin Corporation and now its CEO. *(Heidi Wood 2002)*

During our final stop in Moscow, we enjoyed dinner with Mikhail Gorbachev, during which we discussed the future of American aerospace trade with Russia. Pictured from left to right: Whit Peters, the author, Alexander Vershbow (the American ambassador to Russia), Gorbachev, and his personal translator. (*Family Archives 2002*)

4.
DARK MATTERS

There is no way to put it gently. The twentieth century ended without us knowing the composition of ninety percent of the matter in the universe. What we call "dark matter" emits no light in any form and does not interact with ordinary (household) matter by any known means. Its identity remains a mystery, although its gravity is immediately apparent. In an example from our home galaxy, the outer regions of our Milky Way revolve around the galactic center ten times faster than they otherwise would, were it not for the actions of dark matter. Ordinary matter and dark matter coexist, not in parallel universes, but side by side in the same universe. They feel each other's gravity, but otherwise do not respond to each other's presence.

Perhaps astrophysicists are at the dawn of a new era of scientific discovery, just as the physicists were in 1900. At the time, various loose threads in prevailing theories began to unravel. They eventually unraveled completely, opening the door to entirely new branches of physics. One of them, called quantum mechanics, accurately accounts for nature's behavior on its smallest scales—molecules, atoms, and particles. Nearly a dozen Nobel Prizes were awarded to the scientific leaders of that effort.

Dark matter may be a rich assortment of exotic subatomic particles, as some theories have proposed. But it may be something yet to be imagined. The dark matter dilemma in astrophysics at the dawn of the twenty-first century may force a revolution in our understanding of gravity and (or) matter that rivals the scientific revolutions of the past.

Occasionally, I cannot help but personalize, even personify, dark matter's place in the universe—especially the part about matter and dark matter feeling each other's gravity but not otherwise interacting. This schism came home to me during the summer of 1991, when I attended an annual conference of one of the national physics societies of which I am a member, near Atlanta, Georgia. That fall I would begin my postdoctoral appointment at Princeton. At such conferences, physicists gather from across the country, leaving their academic hamlets, their industrial labs, their particle accelerators, and their government installations to share the latest, yet-to-be-published results on the frontier of human understanding of the universe. During the prescheduled coffee breaks in the common areas of these out-sized hotels, people engage in intense discussions about that day's presentations.

There is something truly comforting at physics conferences. They are places where you feel as if you know people who you have never met before because everyone's life path strongly resembles your own. Surely this is true for all conferences, no matter the field. Among professional physicists, for example, we all got good grades in school (physicists are disproportionately represented among college seniors who graduate magna and summa cum laude). We have all solved the same homework problems in physics classes. We have all read the same books. We wield nearly identical vocabulary sets when describing the physical world. And we have all felt the occasional aspersions cast by pop culture on our intellectual abilities.

By the time of the society banquet, held the last night of the conference, people have loosened up. Discussions are more likely to touch upon personal matters and other things that have nothing to do with the subjects and themes of the conference. By the end of this particular banquet, a dozen of us from several contiguous tables collected the unfinished bottles of wine and retreated to one of those penthouse common rooms on the top floor of the hotel. We talked (and argued) about the sorts of things that the rest of society would surely consider to be geeky and pointless such as why a can of Diet Pepsi floats while a can of regular Pepsi sinks. That one was new to me, although I did have latent memories from the end of long parties where all the ice had melted in the beverage cooler and some soft drinks were floating while others were resting at the bottom.

We lamented the fact that the transporter in the television and film series *Star Trek* does not transport perfectly across space. Apparently, the teleported copy sustains an extremely small but quantifiable level of degradation when compared with the original—a perversely humorous fact that was well known among the Star Trek cognoscenti. The questions to be debated started rolling: How many times could you be transported back and forth between the starship and a planet before you started to look different? What part of your body would change? Was it your DNA? Was it your atomic structure? Or would you one day beam back to the ship without a nose?

We also debated the popularity of the physicist Stephen Hawking, who is known to the public primarily through his bestselling book on the state and fate of the universe, *A Brief History of Time*. Some felt that he has been overrated as a scientist by the public as well as by other scientists. We all agreed that he is a pretty smart guy, and that he is an excellent physicist. But we further agreed that he falls below a dozen other physicists from the twentieth century, most of whom the public has never heard of. We bandied about a short list, including Bohr, de Broglie, Dirac, Eddington, Fermi, Friedmann, Gamov, Gel-Mann, Heisenberg, and Planck. One of my colleagues, a theoretical physicist whose expertise overlaps with that of Hawking, spent some time detailing all the theories that Hawking got wrong. Yes, physicists can be petty and gossipy too.

The evening was rich in the expression of applied mental energy. What else could you have expected among intellectual soul mates, at the end of a full meal, near the end of a full conference, while sipping good wine into the late evening?

Around midnight, during our discussion on momentum transfer in car accidents, one of us mentioned a time when the police stopped him while driving his car. They ordered him from his sports car and conducted a thorough search of his body, the car's cabin, and the trunk before sending him on his way with a hefty ticket. The charge for stopping him was driving twenty miles per hour over the local speed limit. Try as we did, we could not muster sympathy for his case, although a brief discussion of the precision of police radar guns followed. We all agreed that on a straight road, radar guns cannot possibly register your exact speed unless the police officer stands in the middle of the oncoming traffic. If the officer stands anywhere else, the measured speed will be less than your actual speed. So if you were measured to be speeding, you were speeding.

My colleague had other encounters with the law that he shared later that night, but his first started a chain reaction among us. One by one we each recalled multiple incidents of being stopped by the police. None of the accounts were particularly violent or life threatening, although it was easy to extrapolate to highly publicized cases that were. One of my colleagues had been stopped for driving too slowly. He was admiring the local flora as he drove through a New England town in the autumn. Another had been stopped because he was speeding, but only by five miles per hour. He was questioned and then released without getting a ticket. Still another colleague had been stopped and questioned for jogging down the street late at night.

As for me, I had a dozen different encounters to draw from. There was the time I was stopped late at night at an underpass on an empty road in New Jersey for having changed lanes without signaling. The officer told me to get out of my car and questioned me for ten minutes around back with the headlights of his squad car brightly illuminating my face. Is this your car? Yes. Who is the woman in the passenger seat? My wife. Where are you coming from? My parents' house. Where are you going? Home. What do you do for a living? I am an astrophysicist at Princeton University. What's in your trunk? A spare tire, and a lot of other greasy junk. He went on to say that the "real reason" he stopped me was because my car's license plates were much newer and shinier than the seventeen-year-old Ford I was driving. The officer was just making sure that neither the car nor the plates was stolen.

In my other stories, I had been stopped by the police while transporting my home supply of physics textbooks into my newly assigned office in graduate school. They had stopped me at the entrance to the physics building where they asked accusatory questions about what I was doing. This one was complicated because a friend offered to drive me and my boxes to my office. (I had not yet learned to drive.) Her car was registered in her father's name. It was 11:30 PM. Open-topped boxes of graduate math and physics textbooks filled the trunk. And we were transporting them into the building. I wonder how often that scenario shows up in police training tapes.

In that conference hotel room, we exchanged stories about the police for two more hours before retiring to our respective hotel rooms. Being mathematically literate, of course, we looked for "common denominators" among the stories. But we had all driven different cars—some were old, others were new, some were undistinguished, others were high-performance imports. Some police stops were in the daytime, others were at night. Taken one by one, each encounter with the law could be explained as an isolated incident where, in modern times, we all must forfeit some freedoms to ensure a safer society for us all. Taken collectively, however, you would think the cops had a vendetta against physicists because that was the only profile we all had in common. One thing was for sure, the stories were not singular, novel moments playfully recounted. They were common, recurring episodes. How could this assembly of highly educated scientists, each in possession of the PhD—the highest academic degree in the land—be so vulnerable to police inquiry in their lives? Maybe the police cued on something else. Maybe it was the color of our skin. The conference I had been attending was the twenty-third meeting of the National Society of Black Physicists. We were guilty not of DWI (driving while intoxicated), but of other violations none of us knew were on the books: DWB (driving while black), WWB (walking while black), and of course, JBB (just being black).

A year after the conference, Rodney King was pulled from his car by the Los Angeles Police and, while handcuffed, "tasered," and lying facedown on the street, was beaten senseless with night sticks. What sometimes goes unremembered is that the deadly riots that followed in south-central Los Angeles were not triggered by the beating itself but by the subsequent acquittal in the court of law of key participating officers. Upon seeing the now-famous amateur footage of the incident I remembered being surprised not because Rodney King was beaten by the police but because somebody finally caught such an incident on tape.

The next meeting of the National Society of Black Physicists (NSBP), held in Jackson, Mississippi, happened to coincide with those Los Angeles riots. I was scheduled to give the luncheon keynote address on May 1, 1992, on the success or failure of undergraduate physics education in the academic pipeline that leads to the PhD. While watching the helicopter news coverage of the fires and violence that broke out that morning, I had a surreal revelation: the news headlines were dominated by black people rioting and not about black physicists presenting their latest research on the nature of the universe. Of course, by most measures of news priorities, urban riots trump everything else, so I was not surprised. I was simply struck by this juxtaposition of events, which led me to abandon my original keynote address and replace it with ten minutes of reflective observations on NSBP's immeasurable significance to the perception of blacks by whites in America.

During the spring of my sophomore year at Harvard, I was well into the course work of my declared major, taking an (un)healthy dose of physics and math classes as well as the requisite other nonscience courses that a full schedule requires. That year I was also on the university's wrestling team, as second string to a more talented senior in my 190-pound-weight category. One day after practice, we were walking out of the athletic facility when he asked me what I had been up to lately. I replied, "My problem sets are taking nearly all of my time. And I barely have time to sleep or go to the bathroom." Then he asked me what my academic major was. When I told him physics, with a special interest in astrophysics, he paused for a moment, waved his hand in front of my chest, and declared, "Blacks in America do not have the luxury of your intellectual talents being spent on astrophysics."

No wrestling move he had ever put on me was as devastating as those accusatory words. Never before had anyone so casually, yet so succinctly, indicted my life's ambitions.

My wrestling buddy was an economics major and, a month earlier, had been awarded the Rhodes scholarship to Oxford where, upon graduation, he planned to study innovative economic solutions to assist impoverished urban communities. I knew in my mind that I was doing the right thing with my life (whatever the "right thing" meant), but I knew in my heart that he was right. And until I could resolve this inner conflict, I would forever carry a level of surpressed guilt for pursuing my esoteric interests in the universe.

During graduation week of my senior year of college, an article appearing in the *New York Times* broadly profiled the 131 black graduates of my Harvard class of 1,600 people. The *Times* made public for the first time that only 2 of the 131 graduates had plans to continue for advanced academic degrees. I was one of those two. The rest were slated for law school, medical school, business school, or self-employment. (The other "academic" was a friend of mine from the Bronx High School of Science who graduated college in four years with both his bachelor's and master's degrees in history.) Given these data I became further isolated from the brilliant good-deed doers of my generation.

Nine years passed. Having earned my master's degree from the University of Texas at Austin, I spent several more years there before leaving to teach for a year at the University of Maryland and finally transferring my doctoral program to Columbia University. At Columbia, I was well on my way to completing the PhD in astrophysics when I received a phone call at my office from the local affiliate of FOX news. I had already been the department's unofficially designated contact for public and media inquiries about sky phenomena, so this call was not itself unusual—except that it would change my life.

Some explosions reported on the Sun were identified by a recently launched solar satellite, and the FOX news desk wanted to know if everything would be okay in the solar system. After offering my assurances that we would all survive the incident, they invited me to appear in a pretaped interview to convey this information for that evening's broadcast. When I agreed, they sent a car to pick me up. Graduate students are generally not known for fashion or neatness, and I was no exception. Between the phone call and when the car arrived, I ran home, shaved, and put on a jacket and tie. At the television station I was interviewed by the weatherman in a comfortable chair in front of a bookshelf filled with fake, sawed-off books. The interview lasted two minutes, within which I said that explosions on the Sun happen all the time, but especially on eleven-year cycles during "solar maximum" when the Sun's surface is more turbulent than usual. During these times, high doses of charged subatomic particles spew forth from the Sun and fly through interplanetary space. Those particles that head toward Earth deflect toward the poles by the action of Earth's magnetic field. Subsequent collisions of these particles with molecules in Earth's upper atmosphere create a dancing curtain of colors, visible primarily in the arctic regions. These are the famous northern (and southern) lights. I assured the viewers that Earth's atmosphere and magnetic field protect us from these hazards and that people might as well take the opportunity to travel north in search of these displays.

The interview took place at 3:00 PM and was scheduled to air during FOX's six o'clock news. I promptly called everybody I knew and rushed home to watch. That evening, while eating dinner, the segment aired. In the middle of my mashed potatoes, I had an intellectual out-of-body experience. At home I was the general public, yet on the screen before me was a scientific expert on the Sun whose knowledge was sought by the evening news. The expert on television happened to be black. At that moment, the entire fifty-year history of television programming flew past my view. At no place along that timeline could I recall a black person (who is neither an entertainer nor an athlete) being interviewed as an expert on something that had nothing whatever to do with being black. Of course there had been (and continued to be) black experts on television, but they were politicians seeking support and monies for urban programs to help blacks in the ghetto. They were black preachers and other clergy offering spiritual leadership. They were black sociologists analyzing crime and homelessness in the black community. They were black business executives talking about enterprise zones in the most impoverished regions of town. And they were black journalists, writing about black issues.

For the first time in nine years I stood without guilt for following my cosmic dreams. I realized as clear as the crystalline spheres of antiquity that one of the major barriers to successful relations between blacks and whites is the latent supposition that blacks as a group, are just not as smart as whites. This notion runs deep— very deep. It's fed in part by differences in IQ scores and in other standardized exams such as the SATs, where whites score higher than blacks. Its influence is felt in debates on academic tracking, affirmative action (in schools and the workplace), and the international politics of Africa.

The most pervasive expression of the problem is the casually dismissive manner in which many whites treat blacks in society. I have never had an IQ exam, which is possible in this world if you attend only public instead of private schools. I nonetheless know, from reading extensively on the subject, all about them and what they look like. Among the claims of IQ proponents is the fact the single number, your intelligence quotient, is largely inbred and is an indicator of your innate intelligence and your likelihood of succeeding in life. Data show that blacks, on average, score a full standard deviation lower than whites. The prevailing notion is that you cannot substantially increase your IQ at any time, so one might conclude that whites are genetically higher scorers, independent of upbringing, accumulated wealth, or birthright opportunities.

Since humans can get better and better in everything else that matters in the world simply by practicing, I have always questioned the relevance of the IQ exam to one's promise and performance in life. If one's ability to succeed were strongly dependent on a heritable IQ, then why do some whites fear integrated schools? Why the high anxiety and the intense competition that surrounds school choice, from prekindergarten through college? Why the heavy monetary investment in education among those who can afford it? This collective behavior betrays a deep notion that it is wealth and choice of schools, not IQ, that are the most significant factors influencing one's chances of success in life.

Since the adjectives "smart" and "genius" get applied to scientists far more often than to people in other professions, this most fundamental barrier in "race relations" had yet to be crossed. Indeed, the barrier's true nature had yet to be identified.

The incentive to achieve knows no bounds. My father's high school gym instructor singled him out in class as having a body type that would not perform well in track events. My father's muscular build did not fit the lean stereotype of a runner that the instructor had formulated. My father had never run before. But almost out of spite, he went on to become a world-class track star in the 1940s and 1950s—at one time capturing the fifth-fastest time in the world for the 600-yard run. After college, my father continued to run for the New York Pioneer Club, an amateur track organization whose doors were open to blacks and Jews and anybody else who was denied admission to the WASP-only athletic clubs. One of my father's long-time friends and Pioneer Club buddies once competed in a race where he was barely ahead of the number-two runner as they approached the final straightaway. At that moment, the coach of the other runner loudly yelled, "Catch that nigger!" In the world of epithetic utterances, this one ranks among the least intelligent. My father's friend, having overheard the command, declared to himself, "This is one nigger he ain't going to catch" and won the race by an even larger margin.

An academic counterpart to the phrase "Catch that nigger" may be found in my growing collection of scholarly books over the centuries that assert the inferiority of blacks. One of my favorites comes from the 1870 study *Hereditary Genius: An Inquiry into Its Laws and Consequences*, by the English sociobiologist Francis Galton, founder of the Eugenics movement. In the chapter titled "The Comparative Worth of Different Races," he notes:

The number among the negroes of those whom we should call half-witted men, is very large. Every book alluding to Negro Servants in America is full of instances. I was myself much impressed by this fact during my travels in Africa. The mistakes the negroes made in their own matters, which were so childish, stupid, and simpleton-like, as frequently to make me ashamed of my own species.*

---

*Francis Galton, *Hereditary Genius: An Inquiry into Its Laws and Consequences* (New York: D. Appleton, 1870), p. 339.

Whenever I need energy to fight the pressures of society, I just reread one of these passages and, like my father's track buddy, I instantly summon the energy within me to ascend whatever mountain lay before me.

By winning four gold medals and four world records in track and field, Jesse Owens wiped the slate clean of Aryan claims to physical superiority during the 1936 Berlin Olympics. So too will a black American Nobel Laureate (in a category other than peace) forever change the dialog on innate intellectual differences. Who knows when that time will come. In the interim, I play my small part in this journey. I've been interviewed on network television fifty times over the past five years for my expertise on all aspects of modern cosmic discovery—from discoveries in the solar system to theories of the early universe. And I have refused all invitations to speak for black history month on the premise that my expertise is neither seasonal nor occasional. I had finally reconciled my decade of inner conflict. It's not that the plight of the black community cannot afford having me study astrophysics. It's that the plight of the black community cannot afford it if I don't.

My life's goal from the age of nine had always been the PhD in astrophysics. When I finally achieved it, I was determined to share what I had been through, and was given the honor of addressing my fellow doctoral candidates in all disciplines from anthropology to zoology at Columbia University's PhD graduation ceremony.

When the dean first asked me to speak, it occurred to me that I had nothing to say. I could chat about my research, but the audience wouldn't understand the contents of my dissertation any more than I would understand the contents of theirs if it was they who were talking to me. I could expound upon the role of high-level academia in modern society, but you could get that at any convocation or commencement.

My inspiration for the address came from a mountaintop in the Andes Mountains in Chile, where I lived nocturnally for seven days. The trip's purpose was to obtain data on the structure of the galaxy from a location seven thousand feet above sea level at the telescopes of the Cerro Tololo Inter-American Observatory, fifty kilometers from the nearest town. It is there that I obtained nearly all my thesis data, and it is there that I reflected upon my life's path through time and space.

When I was in elementary school in the public schools of New York City, I distinctly remember that it was important for me to be athletic—in particular, to be able to run fast. I was encouraged by all around me. My reward was the respect and admiration of classmates and especially my streetmates.

In junior high school it was important for me, now that I was certified the "fastest on the block," to slam-dunk a basketball. To do this you have to jump high *and* palm the basketball. On April 17, 1973, I was the first in my grade to slam-dunk a basketball. I then asked myself, "Is this all there is to it?" The answer is basically yes, yet one can imagine creative variations such as a 360-degree pirouette in midair preceding the dunk, but you still score only two points.

About the same time, I learned that light, traveling at 186,282 miles per second, moves too slowly to escape from the event horizon of a black hole. This was more astonishing to me than a 360-degree slam-dunk. I soon became scientifically curious and read everything I could find about the universe. I began to see myself as a future scientist—in particular, an astrophysicist. It became a deeply seated dream.

I shortly came to the shattering awareness that few parts of society were prepared to accept my dreams. I wanted to do with my life what people of my skin color were not supposed to do. As an athlete, I did not violate society's expectations since there was adequate precedent for dark-skinned competitors in the Olympics and in professional sports. To be an astrophysicist, however, became a "path of most resistance." I began to wonder whether I originally wanted to be an athlete more from society's interest rather than my own. My brother, Stephen, today a professional artist, could run faster and jump higher than I could. He, too, felt these forces of society.

In high school, nobody probed further about how I became captain of the wrestling team. But when I became editor in chief of my school's annual *Physical Science Journal,* my qualifications were constantly queried. And when I was accepted to the college of my choice, I was continually asked for my SAT scores and grade point average. Indeed, one fellow student, who worked in the office of the guidance counselor, threatened to find the file in the school records to read my scores himself, if I didn't tell.

When I first entered graduate school, before transferring to Columbia, I was eager to pursue my dreams of research astrophysics. But the first comment directed to me in the first minute of the first day, by a faculty member whom I had just met was, "You must join our department basketball team." As the months and years passed, faculty and fellow students, thinking that they were doing me a favor, would suggest alternative careers for me.

"Why don't you become a computer salesman?"

"Why don't you teach at a community college?"

"Why don't you leave astrophysics and academia? You can make much money in industry."

At no time was I perceived as a future colleague, although this privilege was enjoyed by others in graduate school.

When combined with the dozens of times I have been stopped and questioned by the police for going to and from my office after hours, and the hundreds of times I am followed by security guards in department stores, and the countless times people cross the street upon seeing me approach them on the sidewalk, I can summarize my life's path by noting the following: in the perception of society, my athletic talents are genetic; I am a likely mugger-rapist; my academic failures are expected; and my academic successes are attributed to others.

To spend most of my life fighting these attitudes levies an emotional tax that constitutes a form of intellectual emasculation. My Columbia PhD, conferred in 1991, brought the national total of black astrophysicists from six to seven, out of four thousand nationwide. Given what I experienced, I am surprised that many survived.

I eventually learned that you can be ridden only if your back is bent. And, of course, that which doesn't kill you makes you stronger. When I had finally transferred my graduate program to Columbia University, where I was welcomed by the Department of Astronomy, I received a twice-renewed NASA research fellowship, published four research papers, attended four international conferences, had two popular-level books published, was quoted three times in the *New York Times*, appeared twice on network television, and was appointed to a well-respected postdoctoral research position at Princeton University's Department of Astrophysical Sciences.

There are no limits when you are surrounded by people who believe in you, or by people whose expectations are not set by the short-sighted attitudes of society, or by people who help to open doors of opportunity, not close them.

On my college application a questions asks, "What are your goals?" My response was simply, "A PhD in astrophysics," a goal that had been planted within me from when I was nine years old. With the conferral of my Columbia PhD, I had lived and fulfilled my dream, yet I knew my life had just begun and that my struggle would continue.

I was once a family guest on a hilltop wedding reception for my sister-in-law in the farmlands of Washington State. Attending was the usual complement of extended family as well as some neighborhood friends. At dusk on this windless day, as part of the celebration, a small, low-flying crop duster approached the hill and poured bushels of popped corn all over the grass where we were gathered. The corn descended in slow motion, like windblown dandelion seeds. While I was eating the ceremonial corn off the ground (and off people's heads), I wondered to myself whether the popcorn had fallen straight down, or whether it landed forward or backward from the spot where it was released from the airplane. From the point of view of the pilot, of course, all popcorn moves immediately backward from its release point. My question, however, was intended from the point of view of someone standing on the ground.

Fully popped corn has such high air resistance that one might expect them to lose their airplane speed immediately and fall straight to the ground. But the rearward-moving air from the propeller blades might have thrown them backward from their actual drop point in spite of the forward speed of the aircraft. Since I didn't immediately figure out the answer, I decided to ask the question of one of the guests, who was alone and quietly sipping champagne. I think he was an instructor at the local university. Upon hearing my question, he instantly assumed I was ignorant of all matters scientific and described, in a patronizing tone, how the low density of popcorn allows it to encounter very high air resistance upon being released from the airplane. Of course I already knew this. My query had been more subtle. I said to him, "I am not convinced that the corn will have no backward motion after it is dropped. The wash from the propeller may send it backward."

He proceeded to pick up a popped kernel of corn from the grass and impatiently dropped it from his hand to demonstrate his point, as though I were the one in the conversation who was dense. He was surely oblivious to the fact that our conversation was one of the most patronizing I have ever endured. As a matter of social policy, I do not voluntarily convey my formal scientific or educational background in conversations with strangers unless they ask for the information directly. Not knowing my background, he must have thought me to be an ungrateful moron to question his popcorn tutorial.

My father-in-law is an MIT engineer with a scientific pedigree traceable to the post–World War II nuclear arms effort. During the hilltop reception, at a point where his attention was no longer needed by the photographers and by other wedding matters, he walked up to the two of us. I don't know whether my father-in-law had overheard the tone of our banter and felt that I needed to be rescued, but his first comment to me was, "Neil, do you still teach astrophysics at Princeton University?"

This simple nine-word question conveyed reams of data to my patronizing partner that cut through his intellectual aggression. His tone instantly became humble and docile, and he even started to ask scientific questions of me, but he never explicitly apologized for his behavior. We ended up friends by the end of the reception, with him asking about the latest theories on big bang cosmology and on the search for other planets in the galaxy.

I actually find it amusing when people who do not know me, and whom I have never met, assume me to be deeply ignorant. Sometimes their behavior persists even when I make comments that clearly require years of academic study or other intellectual investment. I once walked into a posh wine shop in the Upper West Side of Manhattan in New York City and noticed a bottle of red Bordeaux on the shelf from a particular Chateau and from a particular vintage that I had been in search of for some time. The bottle was seventeen years old, which is not particularly unusual for a fine wine shop. As I reached for the bottle to inspect it, the wine merchant, whom I later learned was also the owner, barked from across the store, "Don't touch the bottle, it's very old and expensive." I replied, "How else am I to decide whether to buy it?" At which point he grudgingly allowed me to lift it off the rack. I then looked through the glass of the wine bottle, toward an incandescent lamp, to judge the wine's color. If it had been stored poorly over the years, the wine's deep, brilliant garnet color would prematurely turn amber and then brown. After I told him that the wine was turning sooner than a Bordeaux of that age and vintage should, he suggested that it was the dust on the bottle that was brown rather than the wine itself.

How many more hints can I give him that I know what I am talking about? After a few more minutes of this charade I politely left the store, went home, and wrote him a letter that I sent via overnight FedEx delivery. After reminding him of our encounter the previous day, my otherwise cordial letter included the following comments:

Dear Mr. Smith,

You must have assumed me to be a fool and yourself to be a wine expert, but clearly the relevant facts suggest just the opposite. Your actions yesterday . . . discredited yourself as an honest wine merchant. Perhaps your other clientele won't know the difference, but if I were you, I would be a little more careful about what you assume to be the background of your customers.

I went on to tell him:

If you cannot rise above your prejudice, then I simply ask to be treated with respect—not because of my extensive wine knowledge, but simply because I was a customer who was prepared to spend money in your store.

If I were of a different character, I could make a lot of money based on this inherent social discrepancy, as others have already done. Depending on the time of year and on the time of day, you can find rows of chess tables near major New York City tourist centers such as Times Square, or near heavily visited public spaces such as Battery Park or Union Square. For $20 or $40 in cash you can place a bet that you will beat the person seated in a game of speed chess. Nearly everyone who is waiting to be challenged is black or otherwise quite urban in dress and manner. The passersby are typically white tourists or businessmen who see this as an opportunity to win some fast cash during their lunch break. They might have been star chess players in their high school or college days, or perhaps they just like the game and can't imagine being beaten by a black person at something that requires smarts and nimble thinking. After all, you are not betting that you will beat him in a slam-dunk contest. In either case, a flippant assumption gets made about the relative intellectual capacity of the passerby and the chess player.

In every case that I have witnessed, the white person lost the game.

In another scenario, I had been trying on shoes in a central New Jersey mall. When nothing suited my tastes I left the store empty-handed, except for my shopping bag from purchases at other stores in the mall. As I passed those metal security loops at the exit door the alarms went off. The security guard stopped me in my tracks and (politely) asked to look in my shopping bag. Meanwhile, the white woman who passed the security loops at the exact moment that I had walked free and clear from the store.

What a brilliant shoplifting scheme, complete with the poetic justice of a store suffering directly from its own prejudice.

Where and when are all these assumptions born? Do people really think that all blacks are criminals and inherently less intelligent (or just stupid) and that whatever status they achieve is the product of affirmative action and of opportunities that they do not deserve? Occasionally, one sees court cases of white students denied admission to one institution or another because minorities gained special access to 10 percent of the slots. One could just as easily interpret these same cases as white people who failed to be selected for 90 percent of the slots because their record was so poor.

I don't suppose I will ever know how far I have gotten in life through the formal or informal application of affirmative action policies. My grades were certainly all over the place throughout my years in school, almost regardless of the course's level of difficulty. As noted earlier, however, they included the highest GPA in the entire seventh grade of a junior high school in Lexington, Massachusetts, a well-to-do suburb of Boston. My grades also include the highest score in the Bronx High School of Science on the junior-year math Regents exam, and the ninety-ninth percentile on the mathematics SAT. And in my adult life I have authored or coauthored seven books. Do I deserve special treatment for the color of my skin? I don't know. But what I do know is that in spite of people assuming that I am intellectually incapable, I have retained enough confidence in myself to treat these encounters as the entertaining side shows that they are.

I am certain, however, that many others do not share this same thickness of skin to withstand the constant onslaught of one's intelligence and ambitions. I occasionally wonder how I have survived it myself.

5.
ROMANCING
THE COSMOS

I ultimately did reach the mountaintop—that coveted destination I had sought ever since my first view of the night sky through binoculars back in elementary school. Actually I reached many mountains, beginning with the McDonald Observatory of the University of Texas, located on Mount Locke in west Texas, one of the most remote (and darkest) regions within the continental United States. My first research paper, published while in graduate school, was based on data I obtained with a collaborator at McDonald Observatory.

Other mountain hideaways include the Hale Observatories of Mount Palomar in Southern California, and the Kitt Peak National Observatories, of Kitt Peak, Arizona. These telescopes happen to be relatively easy to reach. Other sites require extensive travel arrangements. One such trip is to the Cerro Tololo Inter-American Observatory (CTIO) in the Andes Mountains of Chile, where I have conducted more research on the universe than at any other telescope in the world.

I am a city slicker to the bone, but I must confess that mountains are special places. Some of my deepest thoughts and inspiration for life have come to me while on a mountain. The clarity of the air

149

somehow translates into clarity of thought. I suppose I'm not alone here. Mountains have a rich history for inspiring thought and action. What else would drive otherwise rational people to climb mountains for no other reason than just to see what's on the other side. Furthermore, Moses received the Ten Commandments on a mountain, not in a valley. Mohammed was happy to move to the mountain if the mountain would not move to him. Noah parked his ark on a mountain after the waters abated. And in Martin Luther King Jr.'s prophetic speech, delivered on April 3, 1968, the day before he was assassinated, he proclaimed "I've been to the mountain top . . . And I've seen the Promised Land. I may not get there with you. But I want you to know that we as a people will get to the Promised Land."

I conceived of my PhD convocation address, the most important speech of my life, on a mountain. And if the roof of my Skyview apartment building classifies as a mountaintop, then a lot of inspirational stuff happened there, too. But the Cerro Tololo Inter-American Observatory in Chile remains closest to my scientific soul.

The nighttime sky from CTIO in Earth's Southern Hemisphere offers a different assortment and orientation of cosmic objects from the north. In particular, at 30 degrees south latitude—the location of CTIO—the center of the Milky Way galaxy rises at sunset, sets at sunrise, and passes directly overhead at midnight in June. A large part of my research interests focus on the structure of our Milky Way within about three degrees of the galactic center, otherwise known as the galactic bulge, which is a slightly flattened spherical region that is packed with over ten billion stars—about 10 percent of the galaxy's total. When observing the nearest galaxies, one can typically identify only the brightest of its giant stars. The remaining billions blur into puddles of light. For this reason, the ability to observe individual stars in our own galaxy provides a unique platform to understand the structure and formation of all spiral galaxies.

To meet the galactic bulge, one must first submit, half a year in advance, an observing proposal that outlines an idea, defends its worthiness as a scientific project, and describes in detail the requisite hardware needed to achieve the objectives. Observing time is awarded competitively, where the oversubscription rate can approach a factor of five for the largest of a mountain's array of telescopes. Telescope allocation committees parcel out time in blocks as short as two nights, but are typically four to six nights. In the time allocation, no allowance is made for bad weather where bad weather can simply mean overcast skies.

A week or two before the observing run, I prepare detailed coordinates, assemble finding charts, and collect the assorted manuals and notes from previous observing runs that will be useful in real time at the telescope. Then comes the trip. When flying the five thousand miles due south from the New York metropolitan area to Santiago in June, the local time does not change, which is positively no help when your ultimate mission is to be awake at night and asleep during the day. You will never find an astrophysicist who complains about ordinary jet lag because the largest possible jet lag is twelve hours, and this is precisely what you get when you invert your schedule to become nocturnal. In this effort, a time-zone change would assist only the scheduled shift.

The typical pilgrimage requires a 2.5-hour flight to the Miami International Airport, a two-hour layover, a 7.5-hour flight to the Santiago International Airport, a forty-minute (hazardous) taxi ride to the CTIO "Guest House" in downtown Santiago, an eight-hour layover, a twenty-minute (less hazardous) taxi ride to the Santiago bus station, a seven-hour bus ride north—up the coast along the Andes Mountains—to the La Serena administrative headquarters of CTIO, a night at headquarters, and then a 1.5-hour van ride up the Elquí Valley to the summit of Cerro Tololo. The warm clothes I have brought insulate me from the cold of the Chilean mountain winter. I also maintain a keen eye to the sky for the giant South American condors, whose effortless ascent on the mountain thermals portends a night of difficult observing. Once on the mountaintop, I have twenty-four hours to complete the nocturnal inversion before my date with the photons of light from the galactic bulge begins.

Or one can look at it another way. The well-traveled photons began their journey near the center of our galaxy about twenty-six thousand years ago, which renders them contemporaries of Cro-Magnon. My journey, much shorter perhaps, but with no less drama, began three days before. We meet at a detector in the focal plane of the telescope. I can't help contemplating the fate of those photons not collected by the telescope's giant mirror. Imagine a journey of twenty-six thousand light-years only to miss the telescope and slam into the mountainside. Most photons, however, miss Earth completely and are still in motion through interstellar space. But those I collect—those snatched from the photon stream—are what provide the basis for cosmic discovery in my research career.

The moment has arrived. Time is cherished. Clouds are despised. Photons are coveted. The observatory is now my temple—complete with a dome, a telescope, and the dimly lighted control room with its two dozen computer monitors that stream continuously updated information about the telescope, the detector, the object being observed, the ongoing data reductions, and the local weather.

Assisting me in the control room, on one particular trip, is a renowned colleague and friend of mine who is a pure theorist, which simply means he does not necessarily know one end of a telescope from the other. We are two out of three collaborators on a project to obtain original data on the abundance of heavy elements and the velocities through space for thousands of stars. We will use these data to decode some details of the history and structure of the galactic bulge. My theorist colleague had never been to a large optical telescope, so I thought it would be a good idea to haul him all the way to Cerro Tololo. But five hours after he entered the observatory building, central Chile experienced a 6.5-magnitude earthquake; the detector's optics were shaken out of alignment and several hours of data were corrupted. Either the observer gods were upset because a pure theorist entered sacred ground, or the Andes Mountains are geologically active. Regardless, next time I may leave him at home.

Back at Princeton University, the department offices are equipped with powerful computer workstations where we conduct extensive data reduction and analysis. In a manner not unlike the methods by which paleontologists interpret time scales from fossil evidence in sedimentary rock, we infer a history of star formation from the enrichment of heavy elements among its stars. As prescribed by the big bang, the first gas clouds—and the first generation of stars formed from them—were composed of pure hydrogen and helium. Most of the elements heavier than these two in the universe owe their origin to supernovae, titanic explosions of high-mass stars in their death throes. Loaded with heavy elements, ejected matter from supernovae mixed with the gas clouds from which the next generation of stars formed. For each subsequent generation of stars, the total enrichment of heavy elements continued to rise.

While some stars die shortly after they are born, most live for many billions of years. As a result, when we observe the galactic bulge we see a beehive mixture of eons of stellar generations. The number of stars that have few heavy elements when compared with the number that have many heavy elements can help to untangle the history of star formation. And by tagging each star with a velocity in space and a location in the galaxy, we derive useful information about the mass, the gravity, and the origin of the bulge's structure.

To draw scientific conclusions of high confidence requires data of high quality. An excellent night at the telescope requires the very best atmospheric conditions. From the uneven heating and cooling of Earth's surface, however, the lower atmosphere can be a turbulent place of rising and falling air currents. What was good for ascending condors on mountain thermals is bad for astrophysicists. One consequence is that a star's image becomes an undulating blob of light on the detector, which seriously compromises observing efficiency and data quality. For your own safety, do not ever tell an astrophysicist, "I hope all your stars are twinkling."

As you climb through the lower atmosphere the pressure drops exponentially, so that the top of a mountain only seven thousand feet high—the altitude of Cerro Tololo Observatory—sits above nearly 25 percent of Earth's air molecules, with a corresponding 25 percent drop in atmospheric pressure. These observing conditions dramatically improve most astronomical data. A mountain twice this height, such as Mauna Kea in Hawaii (home of many of the world's largest optical telescopes), rises above 40 percent of Earth's atmosphere and is the location of some of the finest ground-based observations ever made.

That mountains tend to be ideal venues for cosmic inquiry did not escape Sir Isaac Newton in his 1704 treatise on optics. He hypothesized:

> If the Theory of making Telescopes could at length be fully brought into Practice, yet there would be certain Bounds beyond which Telescopes could not perform. For the Air through which we look upon the Stars, is in a perpetual Tremor; as may be seen by the . . . twinkling of the fix'd Stars.

Sir Isaac continued with telescopic foresight:

> The only Remedy is a most serene and quiet Air, such as may perhaps be found on the tops of the highest Mountains above the Grosser Clouds.

An even better "Remedy" is found in the well-publicized Hubble Space Telescope, which was lifted into orbit primarily to escape the degraded image quality and poor resolution that the lower atmosphere imposes on observations of all objects.

The thin air has its drawbacks, however. While living nocturnally during the long Chilean winter nights, I must sustain a level of alertness and intellectual intensity that is without counterpart in everyday life. On the mountain, each breath draws one-fourth less oxygen than at sea level yet I am in computer command of millions of dollars worth of high-precision optics and hardware. The stress forces me to reach a self-induced state of cosmic stimulation. Only while observing do I reflect on how many times in a normal day my mind drifts away from peak intensity through built-in mental pauses such as coffee breaks, lunch breaks, mail breaks, and the occasional stare out of my office window.

I end the final night of the observing sessions as I listen to one of those bombastic classical music finales on the observatory dome's CD player. Often the twenty-odd thumps that end the fourth movement of Beethoven's Ninth Symphony do just fine. I close the observatory slit, which generates a sound that, as you might suspect, resonates in the telescope dome with the acoustic richness of a cathedral. Dark time, that most coveted sequence of observing nights where the Moon is near its new phase, ensures that at the end of an observing run of more than four or five days, morning twilight will contain the rising thin crescent moon low on the horizon, framed in the layered colors of the dawn sky. When viewed from a mountaintop, the presunrise horizon light is no less bright than when viewed from sea level, but the surrounding sky that has yet to be absorbed by dawn is much deeper in its darkness. The result is a stirring sweep from the rich remains of the night sky overhead to the radiant twilight on the eastern horizon.

With my little piece of the universe written to a high-capacity data tape in my breast pocket, I now return home with two backup tapes secured—one in my checked luggage, and one left behind on the mountain.

But times change.

During the term of my joint appointment with Princeton University's Department of Astrophysical Sciences, we are part of a consortium of scientists from a half dozen institutions that own and operate a 3.5-meter telescope at Apache Point, New Mexico. Apache Point is the 9,200-foot summit of a cliff face near Sunspot, New Mexico, home of the National Solar Observatory, which is around the corner from the mountain resort town of Cloudcroft, New Mexico. What makes the Apache Point telescope unusual is that it was conceived and constructed to be run remotely over Internet lines from independent control rooms located at each member site. The Princeton control room was carved into a specially outfitted space in the astrophysics department's basement. In principle, the only difference between observing remotely at Apache Point and observing on location at Cerro Tololo is the "length of the wires" that connect to the back of each computer console.

But as efficient as remote observing is, one cannot deny the absence of a mountain's majesty. For better or for worse, I suppose there will come a time when I tell my grand-graduate students: Back in the old days, the data didn't just appear on our doorsteps. We traveled great distances. We ascended great mountains. We met the universe and its photons face-to-face.

The history of discovery in the physical sciences forms a continuous braid, woven of theoretical and experimental triumphs. Occasionally, a scientist is talented at both, but one's formal training is usually either as a theorist or an experimentalist. In the astrophysical sciences, where laboratory tests of cosmic phenomena are few, experimentalists are more accurately described as observers who, more than likely, use mountaintop telescopes.

Observers and theoreticians are fundamentally different. If an observer's data have a history of being flawed (through inferior methodology or because nobody can reproduce the observation), then future data published by that person may be regarded as suspect—especially if the data overthrow well-tested ideas, or hint at brand-new phenomena. Conversely, when armed with pencil and paper and some equations, the theorist can be wrong many times, as long as an interesting path is taken. Interesting paths often contain keys to further discovery.

In pure math, an algebraic equation simply needs to have its left side numerically equal to its right side, and the equation need not relate in any way to the real world. In the physical universe, however, equations connect measured quantities such as temperatures, energy, velocities, and forces. Someone locked away in a closet can therefore deduce all manner of mathematical theorems (if so inclined) but would unlikely walk out as a leading theoretical physicist. Nature has unlimited power of veto on the ideas of physicists, while mathematics is accountable only to its self-contained logic. Behold the primary reason why child prodigies exist among mathematicians but not among physicists.

The mathematics of cosmic discovery contains a language of complicated-looking algebraic equations. Some are beautiful, others are ugly, but they all are nothing more than the mathematical representation of a physical idea. What distinguishes theories rooted in equations from theories rooted in armchair speculation is that the mathematical image of your ideas forces those ideas—and the deductions drawn from them—to be logically constructed. Arguably the most amazing thing about mathematics, which is a pure invention of the human mind, is that it actually works as a tool to help us decode the universe. There was no tablet in the sky that declared the universe to be mathematically describable. We just figured out that it was. Without math, science would not exist as we know it today.

When I first took calculus in eleventh grade at the Bronx High School of Science, I remember seeing columns and columns of esoteric equations that filled the front and rear inside flaps of the textbook. The notation, though elegant, was entirely unfamiliar. Their meaning and purpose was unknown to me. Half the school year would pass before the fog lifted and I learned all about them. They are derivatives and integrals—elegant ways that calculus operates on changing quantities in nature. The calculus I was learning was what Sir Isaac Newton invented to describe why planets orbit the Sun in the shape of ellipses. I felt enlightened, empowered, and energized to learn more and more math so that no part of the physical universe would be out of my reach.

I am convinced that the act of thinking logically cannot possibly be natural to the human mind. If it were, then mathematics would be everybody's easiest course in school and our species would not have taken several millennia to figure out the scientific method. If you fear equations, then you are not alone. In the preface of the best-selling book *A Brief History of Time*, Stephen Hawking reflects on a comment from a publisher friend that for every equation he chose to include, the number of people who would buy the book would be reduced by half. If Hawking had included only ten equations, the publisher expected the readership to drop by a factor of one-half raised to the tenth power, leaving just one-thousandth of the potential readers. *A Brief History of Time* was not published equationless, but it contained many fewer than it could have. As we all know, Hawking wrote what came to be one of the biggest-selling science books of all time.

If the sight of equations upsets you, consider that they are generally no more complicated than anything else you might not understand on first sight. For example, the following equation—known as a Maxwellian distribution of velocities, and named for the famous English physicist James Clerk Maxwell (1831–1879)—contains a healthy assortment of symbols:

$$F(v)\, dv = 4\pi n \left(\frac{m}{2\pi kT}\right)^{3/2} v^2 exp\, (-mv^2/2kT)\, dv$$

Like many important equations that describe the universe, it is a distribution function, which is a slightly more sophisticated version of the bar charts that are common in daily newspapers, such as *USA Today*, that are prone to pictographs. These types of equations tell us how various features of the universe are organized. The Maxwellian distribution of velocities does just that by enabling us to calculate the fraction of all gas molecules that happen to be moving within a designated range of speed.

When applied to the molecular activity within Earth's lower atmosphere, you can use it to calculate the speed with which the largest number of air molecules moves. It's about 450 meters per second. From this speed you can further calculate (using another formula, of course) the speed of sound through air, which is a closely related quantity.

As I acquired knowledge of math and physics through high school, college, and graduate school, the workings of the world around me became more and more transparent. I could understand, describe, and predict phenomena that previously fell out of my reach and out of my grasp.

Learning what these equations do required fifteen seconds of time to read each paragraph that encloses them. Appreciating the full depth and soul of the equations so that you can use them to communicate with others requires extensive study. But the equations of physics are no more cryptic than communication channels found in other disciplines. For instance, nearly everybody knows that deoxyribonucleic acid is DNA, the molecule that encodes the identity of all known forms of life, but years of study would only begin to achieve a full understanding of its function. Or take *Pachycephalosaurus*, which any eight-year-old child knows is a funny-looking dinosaur with a bulbous, knobby head. But understanding its designation as a genus requires some training far beyond the simple memorization of its name. Chemistry is infamous for its cryptic names of things. One of my favorites is oxymetazoline hydrochloride, which happens to be the active ingredient in my twelve-hour nasal spray. It clears my stuffy nose. But beyond that, I'll need to take a course in pharmacology to understand how and why it works in my nasal passages. And the following four lines from the Prologue to Chaucer's *Canterbury Tales,* penned in Middle English, require no small amount of homework to decode and understand:

And smale foweles maken melodye
That slepen al the nyght with open yë
(So priketh hem nature in hir corages);
Thanne longen folk to goon on pilgrimages.

So don't complain about the obscurity of my equations. Besides, unlike other forms of cryptic communication, equations enable us to predict with high precision the nature and behavior of cosmic phenomena. The long history of religious cults that form around those who claim special powers to predict the future alerts us to the fact that your average scientist could create the most devoted cult the world has ever seen. All a scientist needs to do is hide the equations and the methods from view and reveal to the followers only the predictions: The sun will rise tomorrow at 7:02 AM. A comet with two dozen large pieces will slam into Jupiter's atmosphere. The midday sun will be eaten by darkness. The subject would make a good sociology novel.

Some of an equation's obscurity can, of course, be blamed on the presence of unfamiliar symbols. These days, it's hard to find an equation that does not use one or more squiggly letters from a foreign alphabet. And alphabets don't come squigglier than lowercase Greek. In sequence from alpha to omega we have: $\alpha\ \beta\ \gamma\ \delta\ \epsilon\ \zeta\ \eta\ \theta\ \iota\ \kappa\ \lambda\ \mu\ \nu\ \xi\ o\ \pi\ \rho\ \sigma\ \tau\ \upsilon\ \phi\ \chi\ \psi\ \omega$. The most famous among them is probably the letter pi: π. Pi normally represents the exact ratio of a circle's circumference to its diameter, and thus makes cameo appearances in all manner of equations that contain references to geometry—from the area of a circle to the shape of the universe. By the way, you can always remember the formula for the area of a circle because, as the saying goes: pi are not round, pi are squared. In other "words," $A = \pi r^2$. At least half of the lowercase Greek letters are in regular use by astrophysicists and represent selected physical quantities.

We also tap the uppercase letters of the Greek alphabet: $A\ B\ \Gamma\ \Delta\ E\ Z\ H\ \Theta\ I\ K\ \Lambda\ M\ N\ \Xi\ O\ \Pi\ P\ \Sigma\ T\ \Upsilon\ \Phi\ X\ \Psi\ \Omega$, although many resemble letters from our familiar Roman alphabet. In cosmology, the study of the origin and fate of the universe, one of the most widely used symbols is omega: Ω. Defined as the ratio of the actual density of mass in the universe to a "critical" density, its value tells us whether or not our expanding universe will one day recollapse due to the collective gravity of all cosmic matter. Latest data suggest that $\Omega = 1$, suggesting a flat universe that will expand forever, but one that is on the brink of recollapse. Using the lowercase Greek letter rho (ρ) to represent density, the relation reads:

$$\Omega = \rho / \rho_{crit}.$$

Equations are not ideas unto themselves. They are just the symbols that represent ideas. This subtle, but important, distinction enables quadriplegic Stephen Hawking to deduce the nature of the universe in his head, without having to write the equations on a piece of paper.

As serious as equations can be, their world is not entirely humorless. If an equation happens to have too many Greek letters, you have my permission to say "It's Greek to me." And if you are mathematically disinclined, yet you nonetheless want to make a splash at a party of engineers or physicists, I promise that the following riddle is sure to make them all bust open with laughter:

Q: What do you get when you cross a rabbit with an elephant?
A: Rabbit elephant sine theta.

The above riddle is hilarious because there is a mathematical operation called a "cross product," which takes two quantities, each having a magnitude and a direction (such as two velocities or two forces), and multiplies their magnitudes with the sine of theta (θ)—the angle between the directions they point. The sine function is one of those operations in trigonometry that you were certain you would never see again after high school. Mathematically, the cross product reads

$$| \mathbf{A} \times \mathbf{B} | = A \ B \ sine \ \theta,$$

where the flanking | | symbols provides instructions to calculate the magnitude of the result. In an admittedly absurd algebraic substitution, you set $A = rabbit$ and $B = elephant$ and you recover the structure of the original riddle. Every physics and engineering student learns about cross products—as well as other valuable ways to combine physical quantities—no later than the first year in college.

I first saw the following relation on a bathroom wall in my high school.

$$\int e^x = f(u)^n$$

The long and graceful s-shaped symbol was developed by the famous seventeenth-century German mathematician Gottfreid Leibniz as a stylized letter "s" representing a sum. While not constructed to be a bona fide equation, it does beg to be read as "sex = fun." Such was the bathroom humor of my high school.

My vote for the most profound equations ever conceived goes to a set that, as before, bear the name of the English physicist James Clerk Maxwell. Containing a complete description of the behavior and propagation of electromagnetic waves (i.e., light). Maxwell's equations occupy the summit of classical (pre-twentieth-century) physics. I reproduce them below not because I expect you to calculate with them but because they are beautiful and they reveal a remarkable asymmetry in the universe.

$$\nabla \cdot \mathbf{E} = 4\pi\rho$$

$$\nabla \times \mathbf{E} = -\frac{1}{c}\frac{\partial \mathbf{B}}{\partial t}$$

$$\nabla \cdot \mathbf{B} = 0$$

$$\nabla \times \mathbf{B} = \frac{1}{c}\frac{\partial \mathbf{E}}{\partial t} + \frac{4\pi}{c}\mathbf{J}$$

The $\mathbf{E}$ stands for electric field, the $\mathbf{B}$ stands for magnetic field, and the $\mathbf{J}$ stands for a current of moving charges. In what was formerly considered to be two separate notions, both electricity and magnetism were conjoined in Maxwell's equations to represent a single physical entity known as electromagnetism. Notice also the upside-down pyramid (a special operator) accompanied by a dot next to the letter **E** in the first line. This particular equation describes the behavior of an electric field around charged objects. The counterpart equation for magnetic fields appears on the third line. But the equation equals zero. By a little-understood fluke of nature, the universe contains isolatable electrical charges (pluses and minuses) but no isolatable magnetic charges. What this means is that north poles of magnets always come attached to south poles. Try it. Go home and smash a magnet into smithereens. Each piece will spontaneously become an N-S magnet, no matter how small or large the fragments are. In physics vernacular, the universe contains no monopoles (as revealed in Maxwell's equations). This remains one of the great mysteries of the cosmos.

If you want to know more about Maxwell's equations, they require a background in vector calculus and electrodynamics, which I will not introduce at this time.

Some equations are symmetric and relatively simple, depending on the coordinate system in which they are constructed. Expressed in familiar x, y, and z coordinates, a widely used equation to probe the spatial shape of many things (including the force of gravity), is called the Laplacian operator, named for the brilliant French mathematician Pierre-Simon Laplace (1749–1827):

$$\nabla^2 = \frac{\partial^2}{\partial x^2} + \frac{\partial^2}{\partial y^2} + \frac{\partial^2}{\partial z^2} \ .$$

As a smooth operator, the equation acts as though it were a machine in an assembly line. You feed it a mathematical function and out comes a representation of that function's behavior in three-dimensional space. In many cases, the Laplacian operator is easier to use when you transform x, y, and z into the spherical coordinates of r, θ, and ϕ, which is the natural coordinate system for spherical objects such as stars and galaxy haloes. But it now takes on an intimidating air that has made strong men weep:

$$\nabla^2 = \frac{1}{r^2} \frac{\partial}{\partial r} \left(r^2 \frac{\partial}{\partial r} \right) + \frac{1}{r^2 sin\theta} \frac{\partial}{\partial \theta} \left(\sin\theta \frac{\partial}{\partial \theta} \right) + \frac{1}{r^2 \sin^2\theta} \frac{\partial^2}{\partial \phi^2} .$$

At the Bronx High School of Science, the number of mathematical functions on your pocket calculator earned you more popularity credits than whether you were a star athlete. Immediately after learning of Maxwell's equations, one classmate of mine, Franck Larece, fantasized about a set of mathematical relations that would one day be known as the "Larece equations." Having already learned about Laplace and his equally brilliant contemporary, Joseph-Louis Lagrange (1736–1813), for me it was not a stretch to imagine the name Larece among them.

If your name becomes associated with fertile equations then you become forever linked to continuing pathways of discovery. For example, after some additional mathematical tools were developed by Laplace, Newton's equations of gravity enabled one to infer the existence of a theretofore undiscovered planet in the outer solar system whose gravity tugged on Uranus's orbit. Sure enough, the planet Neptune was discovered just about where it was predicted to be.

Mercury's orbit also behaves in ways that do not strictly follow the predictions of Newton's laws. Mercury's closest approach to the Sun in its oval orbit predictably shifts over time, through the combined tugs of all other sources of gravity in the solar system. However, the observed shift was more than could be credited to Newton's laws once all known sources of gravity, such as all the rest of the planets, were reconciled. After the triumphant discovery of Neptune, astronomers were armed and ready. The fellow who predicted the existence and location of Neptune in 1846, Urbain-Jean-Joseph Leverrier (1811–1877), took on the task. Wasting no time, Leverrier proposed in 1846 a brand-new planet, Vulcan. Named for the Roman god of fire, Vulcan orbited close to the Sun, providing a gravitational tug on Mercury with just enough force to account for the deviations from Newton's laws. Never mind the fact that such a planet could have (and would have) been detected during the count-less total solar eclipses throughout recorded history, Vulcan lived in and among Newton's equations of gravity for seventy years.

When Einstein published in 1916 the general theory of relativity (the modern theory of gravity), his equations showed that in the vicinity of strong sources of gravity, Newton's laws do not provide an accurate description of the behavior of matter. The disturbed fabric of space and time alters what one would expect from Newton's laws alone. Sure enough, the deviation in Mercury's closest approach to the Sun was fully accounted for within Einstein's new theories. In the first case, unexplained planetary perturbations led to the predic-tion and discovery of a new planet. In the second case, unexplained planetary perturbations led to new laws of physics. Such are the schizophrenic paths that lie before the research scientist.

In the history of the physical sciences, when successful theories are supplanted by ones that are more complete, the previous theories (and their attendant equations) don't all of a sudden become ineffec-tive. The genetic links are in place; Einstein's equations look exactly like Newton's equations when you plug in slow speeds and weak gravity. And Newton's equations can be stripped down to look exactly like Johannes Kepler's descriptive laws of planetary motion.

To solve for the unknown quantities within an equation using pencil and paper asserts the same level of noble solitude as writing a letter with a quill pen by candlelight. You become absorbed by the task, which, for complex equations, can last many hours. While en route to the mathematical solution, you forsake food, personal hygiene, and the measurement of time. I have found that when calculating what no one has calculated before, like my observing sessions on the mountain, my mental acuity peaks. Ironically, these are the times that I would flunk the reality check normally reserved for mental patients and dazed boxers: What is your name? What day is it? Who is the president of the United States? During intense computational moments, I do not remember, I do not know, and I do not care. I am at peace with my equations as I connect to the cosmic engines that drive our universe.

6.

THE END
OF THE WORLD

For some people, meteorites are trophies, to be cherished and displayed. For me, they are also harbingers of doom and disaster. Consider that the slowest speed a large asteroid can impact Earth is about six or seven miles per second. Imagine getting hit by my overpriced objet d'art moving that fast. You would be squashed like a bug. Imagine one the size of a beach ball. It would obliterate a four-bedroom home. Imagine one a few miles across. It would alter Earth's ecosystem and render extinct the majority of Earth's land species. That's what meteorites mean to me, and it's what they should mean to you because the chances that both of our tombstones will read "killed by asteroid" are about the same for "killed in an airplane crash."

About two dozen people have been killed by falling asteroids in the past four hundred years, but thousands have died in crashes during the relatively brief history of passenger air travel. The impact record shows that by the end of 10 million years, when the sum of all airplane crashes has killed a billion people (assuming a conservative death-by-airplane rate of a hundred per year), an asteroid is likely to have hit Earth with enough energy to kill a billion people. What confuses the interpretation of your chances of death is that while airplanes kill people a few at a time, our asteroid might not kill anybody for millions of years. But when it hits, it will take out hundreds of millions of people instantaneously and many more hundreds of millions in the wake of global climatic upheaval.

The combined asteroid and comet impact rate in the early solar system was frighteningly high. Theories of planet formation show that chemically rich gas condenses to form molecules, then particles of dust, then rocks and ice. Thereafter, it's a shooting gallery. Collisions serve as a means for chemical and gravitational forces to bind smaller objects into larger ones. Those objects that, by chance, accreted slightly more mass than average will have slightly higher gravity and attract other objects even more. As accretion continues, gravity eventually shapes blobs into spheres and planets are born. The most massive planets had sufficient gravity to retain gaseous envelopes. All planets continue to accrete for the rest of their days, although at a significantly lower rate than when formed.

Still, there remain billions (possibly trillions) of comets in the extreme outer solar system, orbiting up to a thousand times the size of Pluto's orbit. They are all are susceptible to gravitational nudges from passing stars and interstellar clouds that set them on their long journey inward toward the Sun. Solar system leftovers also include short-period comets, of which two dozen are known to cross Earth's orbit, and thousands of cataloged asteroids, of which at least a hundred do the same.

On the return trip across the country from my summer at Camp Uraniborg we took a detour to visit Meteor Crater in Arizona. The juxtaposition of appearance with accurate knowledge can be the most humbling force on the human soul. On first sight, the crater is simply an enormous hole in the ground—fourteen football fields across and deep enough to bury a sixty-story building. With the Grand Canyon a few hundred miles away, Arizona is no stranger to holes in the ground. But to carve the Grand Canyon, Earth required millions of years. To excavate Meteor Crater, the universe, using a sixty-thousand-ton asteroid traveling upward of twenty miles per second, required a fraction of a second. No offense to Grand Canyon lovers, but for my money, Meteor Crater is the most amazing natural landmark in the world.

The polite (and scientifically accurate) word for asteroid impacts is "accretion." I happen to prefer "species-killing, ecosystem-destroying event." But from the point of view of solar system history, the terms are the same. We cannot simultaneously be happy we live on a planet; happy that our planet is chemically rich; and happy we are not dinosaurs; yet resent the risk of planet-wide catastrophe. Some of the energy of an asteroid collision with Earth gets deposited into our atmosphere through friction and an airburst of shock waves. Sonic booms are shock waves too, but airplanes typically make them by traveling at speeds anywhere between one and three times the speed of sound. The worst damage a sonic boom might do is jiggle the dishes in your cabinet. But with speeds upward of forty-five thousand miles per hour—nearly seventy times the speed of sound—the shock waves from your average collision between an asteroid and Earth can be devastating.

If the asteroid (or comet) is large enough to survive its own shock waves, the rest of its energy is deposited on Earth in an explosive event that heats the ground and blows a crater that can measure twenty times the diameter of the original object. If many impactors strike with little time between each event, then Earth's surface would not have enough time to cool between impacts. We infer from the pristine cratering record on the surface of the Moon (our nearest neighbor in space) that Earth experienced an era of heavy bombardment between 4.6 billion and 4 billion years ago. The oldest fossil evidence for life on Earth dates from about 3.8 billion years ago. Before that, Earth's surface was unrelentingly sterilized. The formation of complex molecules, and thus life, was inhibited, although all the basic ingredients were being delivered nonetheless. An often-quoted figure for life to emerge is 800 million years (4.6 billion minus 3.8 billion equals 800 million). But to be fair to organic chemistry, you must first subtract all the time

Earth's surface was forbiddingly hot. That leaves a mere two hundred million years over which life emerged from a rich chemical soup, which, as does all good soups, includes water.

Yes, much of the water you drink each day was delivered to Earth by comets more than 4 billion years ago. But not all space debris are leftovers from the beginning of the solar system. Earth has been hit at least a dozen times by rocks ejected from Mars, and we've been hit countless more times by rocks ejected from the Moon. Ejection occurs when impactors carry so much energy that smaller rocks near the impact zone are thrust upward with sufficient speed to escape the gravitational grip of the planet. Afterward, the rocks mind their own ballistic business in orbit around the Sun until they slam into something. The most famous of the Mars rocks is the first meteorite found near the Alan Hills section of Antarctica in 1984. Officially known by its coded, though sensible, abbreviation, ALH 84001, this meteorite contains tantalizing, though circumstantial, evidence that simple life on the red planet thrived a billion years ago. As noted earlier, a frenzy of media attention greeted this announcement, made in 1996 by a team led by NASA scientists. Mars has boundless "geo"logical evidence for a history of running water that includes dried riverbeds, river deltas, and flood plains. Since liquid water is crucial to the survival of life as we know it, the possibility of life on Mars does not stretch scientific credulity. The fun part comes when you speculate whether life arose on Mars first, was blasted off its surface as the solar system's first bacterial astronaut, and then arrived to jump-start Earth's own evolution of life. There's even a word for the process: panspermia. Maybe we are all Martians.

The claims for life on ALH 84001 remain controversial, as are most claims on the frontier of cosmic discovery. All the more reason to get up and go to Mars and get more data. In the meantime, matter is far more likely to travel all by itself from Mars to Earth than vice versa. Escaping Earth's gravity requires over two and a half times the energy than that required to leave Mars. Furthermore,

Earth's atmosphere is about a hundred times denser. Air resistance on Earth (relative to Mars) is formidable. Bacteria would have to be hardy indeed to survive the several million years of interplanetary wanderings before landing on Earth. Fortunately, there is no shortage of liquid water and rich chemistry on Earth, so we do not require theories of panspermia to explain the origin of life as we know it, even if we still cannot explain it.

Ironically, we can (and do) blame impacts for major episodes of extinction in the fossil record. Seventy percent of Earth's surface is water and over 99 percent is uninhabited, so you would expect nearly all impactors to hit either the ocean or desolate regions on Earth's surface. So why do movie meteors have such good aim? Especially egregious examples include the blockbuster *Armageddon*, where an incoming meteor decapitates New York City's Chrysler Building, and the 1997 television miniseries *Asteroid*, where a meteor squarely hits a dam in Kansas and floods the nearby town. If you are a movie producer, you can still have a grand time destroying all life on Earth. All you need to do is have the asteroid hit the ocean and have the impact induce global tsunamis that wash down the drain the world's coastal cities.

Some famous impact sites in the world include the 1908 explosion near the Tunguska River in Siberia, which felled thousands of square kilometers of trees and incinerated the three hundred square kilometers that encircled ground zero. The impactor was likely a sixty-meter stony meteorite (about the size of a twenty-story building) that exploded in midair, thus leaving no crater. Collisions of this magnitude happen, on average, every couple of centuries. The two-hundred-kilometer-diameter Chicxulub Crater in the Yucatan, Mexico, is likely to have been left by a ten-kilometer asteroid. With an impact energy 5 billion times greater than the atomic bombs exploded in World War II, such a collision might occur about once in a hundred million years. The Chicxulub Crater is dated from 65 million years ago, and there hasn't been one of its magnitude since. Coincidentally, at about the same time, *Tyran-*

nosaurus rex and friends became extinct, enabling mammals to evolve into something more ambitious than tree shrews.

Those paleontologists and geologists who remain in denial of the role of cosmic impacts in the extinction record of Earth's species must figure out what else to do with the deposit of energy being delivered to Earth from space. The range of energies varies astronomically. Most impactors with less than about ten megatons of energy will explode in the atmosphere and leave no trace of a crater. The few that survive in one piece to leave a crater are likely to be iron based.

Fortunately, among the population of Earth-crossing asteroids, we have a chance at cataloging everything larger than about a kilometer—the size that begins to wreak global catastrophe. An early-warning and defense system to protect the human species from these impactors is a realistic goal. Unfortunately, objects smaller than about a kilometer do not reflect enough light to be reliably detected and tracked. These can hit us without notice, or they can hit with notice that is much too short for us do anything about it. The bright side of this news is that while they have enough energy to create local catastrophe by incinerating entire nations, they will not put the human species at risk of extinction. Have a nice day.

The more I study the risks of impacts, the more tentative life on Earth feels to me. Perhaps I know too much to be calm. In the 1998 disaster film *Deep Impact*, a comet does hit the Atlantic Ocean (instead of a landmarked building in a famous city) and spews forth a tidal wave that wipes out the coastal cities of North America, especially New York City. I saw the building in which I currently reside topple like a domino against other buildings in lower Manhattan, as the wall of water plowed through the city and up the Hudson River valley. We, as a species, are utterly helpless in the face of common disasters such as tornadoes, hurricanes, volcanoes, earthquakes, and tsunamis. We can neither control them nor stop them. Yet, the worst of them pale when compared with the devastation a killer asteroid can bring.

Of course, Earth is not the only rocky planet at risk of impacts. Mercury has a cratered face that, to a casual observer, looks just like the Moon. Recent radio topography of cloud-enshrouded Venus shows no shortage of craters either. And Mars, with its historically active geology, reveals large craters that were recently formed.

Earth's fossil record teems with extinct species—life-forms that had thrived far longer than the current Earth tenure of *Homo sapiens*. Dinosaurs are in this list. What defense do we have against such formidable impact energies? The battle cry of those with no war to fight is "blow them out of the sky with nuclear weapons." True, the most efficient package of destructive energy ever conceived by humans is nuclear power. A direct hit on an incoming asteroid might explode it into enough small pieces to reduce the impact danger to a harmless, though spectacular, meteor shower. (In empty space, where there is no air, there can be no shock waves, so a nuclear warhead must actually make contact with the asteroid to do damage.)

Another method engages those radiation-intensive neutron bombs (you remember—they were the variety of bombs that killed people but left the buildings standing) in a way that the high-energy neutron bath heats one side of the asteroid to sufficient temperature that material spews forth and the asteroid recoils out of the collision path. A kinder, gentler method is to nudge the asteroid out of harm's way with slow but steady rockets somehow attached to one side. If you do this early enough, then only a small nudge will be required using conventional chemical fuels. If we cataloged every kilometer-sized (and larger) object whose orbit intersects Earth's, then a detailed computer calculation would enable us to predict a catastrophic collision hundreds, and even thousands, of orbits in the future, granting Earthlings sufficient time to mount an appropriate defense. But our list of potential killer impactors is woefully incomplete and our ability to predict the behavior of objects much farther into the future (for millions and billions of orbits) is severely compromised by the onset of orbital chaos.

Should we build high-tech missiles that live in silos somewhere awaiting their call to defend the human species? We would first need that detailed inventory of the orbits for all objects that pose a risk to life on Earth. The number of people in the world engaged in this search totals one or two dozen. How long into the future are you willing to protect *Homo sapiens* on Earth? Before you answer that question, take a detour to Arizona's Meteor Crater during your next vacation.

Sometimes it seems that everybody is trying to tell you when and how the "world" is supposed to end. Some scenarios are more familiar than others. Those that are widely discussed in the media include rampant infectious disease, nuclear war, environmental decay, and of course collisions with asteroids or comets. While different in origin, each can induce the end of human species (and perhaps other selected life-forms) on Earth. Implicit in clichéd slogans such as "Save the Earth" is the egocentric call to save life on Earth, not the planet itself.

In fact, humans cannot really save Earth, nor can we really kill Earth. Earth will remain in happy orbit around the Sun, along with its planetary brethren, long after *Homo sapiens* has become extinct by whatever cause. But there are less familiar, though just as real, end-of-world scenarios that jeopardize our temperate planet in its stable orbit around the Sun. I offer these prognostications not because humans are likely to live long enough to observe them, but because the tools of astrophysics enable me to calculate them. Three that come to mind are the death of the Sun, the impending collision between our Milky Way galaxy and the Andromeda galaxy, and the death of the universe, about which the community of astrophysicists has recently achieved consensus.

These scenarios of catastrophe do not worry me on a day-to-day basis because they are slow and steady. But I dream about them and how spectacular they would look if you could speed up time. Computer models of stellar evolution are akin to actuarial tables. They indicate a healthy life expectancy of 10 billion years for our Sun. At an estimated age of 5 billion years, the Sun has another 5 billion years of relatively stable energy output. By then, if we have not figured out a way to leave Earth, then we will bear witness to a remarkable evolutionary change in our host star as it runs out of fuel.

The Sun owes its stability to the controlled fusion of hydrogen into helium in its 15-million-degree core. The gravity that wants to collapse the star is held in balance by the outward gas pressure that is sustained by the fusion. While more than 90 percent of the Sun's atoms are hydrogen, the ones that matter are those that reside in the core. When the core is exhausted of its hydrogen, the Sun is left with a central ball of helium atoms that require a higher temperature than does hydrogen to fuse into heavier elements. Now out of balance, gravity wins, the inner regions of the star collapse, and the central temperature rises through 100 million degrees, which triggers the fusion of helium into carbon.

Along the way, the Sun's luminosity grows astronomically, which forces its outer layers to expand to bulbous proportions, engulfing the orbits of Mercury and Venus. Eventually, the Sun will swell to occupy the entire sky as its expansion nearly subsumes the orbit of Earth. This would be bad. The temperature on Earth will rise until it equals the 3,000-degree rarefied outer layers of the expanded Sun. Our atmosphere will evaporate away into interplanetary space and the oceans will boil off as Earth becomes a red-hot charred ember orbiting deep within the Sun. Eventually, the Sun will cease all nuclear fusion, lose its spherical, tenuous, gaseous envelope, and expose its dying central core. Scenarios such as these will one day force manned space travel to become a global priority.

In my first sky show as director of the Hayden Planetarium, I wrote a script called "Cosmic Mind Bogglers." It included a catastrophe or two, just as I had dreamed them to be. For one of the sequences the sun becomes a red giant as it slowly swells to fill the dome of the sky theater. The event is accompanied by an intensely ominous musical track. It must have worked because during the month after its premiere, I promptly received dozens of letters from angry parents whose children could not sleep at night for fear of the Sun's fate.

The kids must not have paid attention to the part where the show's narrator says, "Five billion years from now. . . ."

Not long after the Sun toasts Earth, the Milky Way will encounter some problems of its own. Of the hundreds of thousands of galaxies whose velocity relative to the Milky Way has been measured, only a few are moving toward us. All the rest are moving away at a speed directly related to their distances from us. Discovered in the 1920s by Edwin Hubble (after whom the Hubble Space Telescope was named), the general recession of galaxies is the observational signature of our expanding universe. The Milky Way and the three-hundred-billion-star Andromeda galaxy are close enough to each other that the effect of the expanding universe is negligible. We happen to be drifting toward each other at about 100 kilometers per second (a quarter million miles per hour). If our (unknown) sideways motion is small, then at this rate, the 2.4 million light-year distance that separates us will shrink to zero within about 7 billion years.

Interstellar space is so vast that there is no need to fear whether stars in the Andromeda galaxy will accidentally slam into the Sun. During the galaxy-galaxy encounter, which would be a spectacular sight from a safe distance, stars are likely to pass each other by. But the event would not be worry free. Some of Andromeda's stars are likely to swing close enough to our solar system to influence the orbit of the planets and of the hundreds of billions of resident comets. For example, close stellar flybys can throw one's gravitational allegiance into question. Computer simulations commonly show that the planets are either stolen by the interloper in a "flyby looting" or they become unbound and are flung forth into interplanetary space.

Remember how choosy Goldilocks was with other people's porridge? If we are stolen by the gravity of another star, there is no guarantee that our newfound orbit will be at the right distance to sustain liquid water on Earth's surface—a condition generally agreed to be a prerequisite to sustaining life as we know it. If Earth orbits too close, its water supply evaporates. And if Earth orbits too far, its water supply freezes solid.

By some miracle of future technology, if Earth inhabitants had managed to prolong the life of the Sun, then these efforts will be rendered irrelevant when Earth is flung into space. The absence of a nearby energy source will allow Earth's surface temperature to drop swiftly to hundreds of degrees below zero Fahrenheit. This would also be bad. Our cherished atmosphere of nitrogen and oxygen and other gases would first liquefy and then freeze solid, encrusting the Earth like icing on a cake. We would freeze before we had a chance to starve. The last surviving life on Earth would be those privileged organisms that had evolved to rely not on the Sun's energy but on (what will then be) weak geochemical and geothermal sources, where the heat of Earth's interior emerges from the crust. At the moment, humans are not among them. There will be, of course, other planets that we can visit in orbit around healthy stars in other galaxies.

Even if you manage to stay safe under water, and evolve to dine upon tasty tubeworms at the midocean vents, the long-term fate of the cosmos cannot be postponed or avoided. No matter where you hide, you will be part of a universe that inexorably marches toward a peculiar oblivion. The latest and best evidence available on the space density of matter and the expansion rate of the universe suggest that we are on a one-way trip: the collective gravity of everything in the universe is insufficient to halt and reverse the cosmic expansion.

◆◇◆

Currently, the most successful description of the universe and its origin combines the big bang with our modern understanding of gravity, derived from Einstein's general theory of relativity. The early universe was a trillion-degree maelstrom of matter mixed with energy, affectionately known as the primordial soup. During the 14-billion-year expansion that followed, the background temperature of the universe has dropped to a mere 3 degrees on the absolute (Kelvin) temperature scale. As the universe continues to expand, this temperature will continue to approach zero.

Such a low background temperature does not directly affect us on Earth because our Sun (normally) grants us a cozy life. But as each generation of stars is born from the interstellar gas clouds of the galaxy, less and less gas remains to compose the next generation of stars. Eventually the gas supply will run out, as it already has in nearly half the galaxies in the universe. The small fraction of stars with the highest mass collapse completely, never to be seen again. Some stars end their lives by blowing their guts across the galaxy in a supernova explosion. This returned gas can then be tapped for the next generation. But the majority of stars—the Sun included—ultimately exhaust the fuel at their cores and, after the bulbous giant phase, collapse to form a compact orb of matter that radiates its feeble leftover heat to the frigid universe.

The complete list of corpses may sound familiar: black holes, neutron stars (pulsars), white dwarfs, and even brown dwarfs are each a dead end on the evolutionary tree of stars. What they each have in common is an eternal lock on cosmic construction materials. In other words, if stars burn out and no new ones are formed to replace them, then the universe will eventually contain no living stars.

How about Earth? We rely on the Sun for a daily infusion of energy to sustain life. If the Sun and the energy from all other stars were cut off from us, then mechanical and chemical processes (life included) on and within Earth would "wind down." Eventually, the energy of all motion gets lost to friction and the system reaches a single uniform temperature. This would really be bad. The starless Earth will lie naked in the presence of the frozen background of the expanding universe. The temperature on Earth will drop the way a freshly baked pie cools on a windowsill. Yet Earth is not alone in this fate. Trillions of years into the future, when all stars are gone, and every process in every nook and cranny of the expanding universe has wound down, all parts of the cosmos will cool to the same temperature as the ever-cooling background. At that time, space travel will no longer provide refuge. Even hell will have frozen over. We may then declare that the universe has died—not with a bang, but with a whimper.

7.
GOD AND THE ASTRONOMERS

F or nearly every public lecture that I give on the universe, I try to reserve adequate time for questions at the end. The progression of subjects raised is predictable. First, the questions relate directly to the lecture. They next migrate to sexy astrophysical subjects such as black holes, quasars, and the big bang. If I have enough time at the end to answer an unlimited number of questions, the subject eventually reaches God. Typical questions include, "Do scientists believe in God?" "Do you believe in God?" "Do your studies in astrophysics make you more or less religious?"

Publishers have come to learn that there is a lot of money in God, and the "greatest story ever told," especially when the author is a scientist and when the book title includes the juxtaposition of scientific and religious themes. Successful books include Robert Jastrow's *God and the Astronomers*, Leon M. Lederman's *The God Particle*, Frank J. Tipler's *The Physics of Immortality: Modern Cosmology, God, and the Resurrection of the Dead,* and Paul Davies's two works, *God and the New Physics* and *The Mind of God: The Scientific Basis for a Rational World*. Each author is either an accomplished physicist or astronomer and, while the books are not strictly religious, they allow the reader to bring God into conversations about astrophysics. Even the late Stephen Jay

Gould, a Darwinian pit bull and devout agnostic, has joined the title parade with his recent work, *Rock of Ages: Science and Religion in the Fullness of Life*. The commercial success of these books indicates that there is a hungry audience who see answers that bridge the chasm between science and religion.

Journalists are not immune from this movement. When the structure within the cosmic microwave background radiation was discovered by the satellite known as the Cosmic Background Explorer, the principal investigator of the project tried to impress upon the media the significance of the result to modern cosmology. He simply said, "If you are religious, it's like seeing God." Not losing an opportunity to quote a scientist who invokes the name of God, the press swiftly misquoted the statement with banner headlines that blared, "Astronomers Discover God" and "Astronomers See the Face of God."

After the publication of *The Physics of Immortality*, which explored whether the laws of physics could allow you and your soul to exist long after you are gone from this world, Frank J. Tipler's book tour included many well-paid lectures to Protestant religious groups. This science-God movement has further blossomed in recent years with efforts made by Sir John Marks Templeton, the wealthy founder of the Templeton investment fund, to find harmony and consilience between science and religion. Apart from sponsoring workshops and conferences on the subject, Templeton's annual religion award, with a cash value rivaling that of the Nobel Prize, has recently been won by several prolific religion-friendly scientists.

Let there be no doubt that as they are currently practiced, science and religion enjoy no common ground. As was thoroughly documented in the nineteenth-century tome, *A History of the Warfare of Science with Theology in Christendom*, by the historian and one-time president of Cornell University Andrew D. White, history reveals a long and combative relationship between religion and science, depending on who was in control of society at the time. The claims of science rely on experimental verification, while the claims of religions rely on faith. These are irreconcilable approaches to knowing, which ensures an eternity of debate wherever and whenever the two camps meet. Just as in hostage negotiations, it's probably best to keep both sides talking to each other. But the schism did not come about for want of earlier attempts to bring the two sides together.

Great scientific minds, from Claudius Ptolemy of the second century to Isaac Newton of the seventeenth, invested their formidable intellects in attempts to deduce the nature of the universe from the statements and philosophies contained in religious writings. Indeed, by the time of his death in 1727, Newton had penned more words about God and religion than about the laws of physics, all in a futile attempt to use biblical chronology to understand and predict events in the natural world. Had any of these efforts worked, science and religion today might be one and the same.

But they are not.

The argument is simple. I have yet to see a successful prediction about the physical world that was inferred or extrapolated from the information content of any religious document. Indeed, I can make an even stronger statement. Whenever people have used religious documents to make accurate predictions about our base knowledge of the physical world, they have been famously wrong. Note that a scientific prediction, which is a precise statement about the untested behavior of objects or phenomena in the natural world, should be logged *before* the event takes place. When your model predicts something only after it has happened, then you have instead made a "postdiction." Postdictions comprise the backbone of most creation myths and, of course, of the "Just So" stories of Rudyard Kipling, where explanations of everyday phenomena explain what is already known. In the business of science, however, a thousand postdictions are hardly worth a single successful prediction.

Topping the list of failed predictions are the perennial claims about when the world will end, none of which have yet proven true. But other claims and predictions have actually stalled or reversed the progress of science. We find a leading example in the trial of

Galileo (which gets my vote for the trial of the millennium), where he showed the universe to be fundamentally different from the dominant views of the Catholic Church. In all fairness to the Inquisition, however, an Earth-centered universe made a lot of sense observationally. With a full complement of epicycles to explain the peculiar motions of the planets against the background stars, the time-honored, Earth-centered model had conflicted with no known observations. This remained true long after Copernicus introduced his Sun-centered model of the universe a century earlier.

The Earth-centric model was also aligned with the teachings of the Catholic Church and prevailing interpretations of the Bible, wherein Earth is unambiguously created before the Sun and the Moon as described in the first several verses of Genesis. If you were created first, then you must be in the center of all motion. Where else could you be? Furthermore, the Sun and Moon themselves were also presumed to be smooth orbs. Why would a perfect, omniscient deity create anything else?

All this changed, of course, with the invention of the telescope and Galileo's observations of the heavens. The new optical device revealed aspects of the cosmos that strongly conflicted with people's conceptions of an Earth-centered, blemish-free, divine universe. The Moon's surface was bumpy and rocky. The Sun's surface had spots that moved across its surface. Jupiter had moons of its own that orbited Jupiter and not Earth. Venus went through phases, just like the Moon. For his radical discoveries, which shook the Christian world, Galileo's books were banned, and he was put on trial, found guilty of heresy, and sentenced to house arrest. This was benign punishment when one considers what happened to the monk Giordano Bruno. A few decades earlier, Bruno had been found guilty of heresy and burned at the stake for suggesting that Earth may not be the only place in the universe that harbors life.

I do not mean to imply that competent scientists, soundly conducting the scientific method, have not also been famously wrong. They have. Most scientific claims made on the frontier will ultimately be disproved, usually with the arrival of more or better data. But this scientific method, which allows for expeditions down intellectual dead ends, also promotes ideas, models, and predictive theories, and that can be spectacularly correct. No other enterprise in the history of human thought has been as successful at decoding the ways and means of the universe.

Scientists are occasionally accused by others of being closed minded or stubborn. Often people make such accusations when they see scientists swiftly discount astrology, the paranormal, Sasquatch sightings, and other areas of human interest that routinely fail double-blind tests or that possess a dearth of reliable evidence. But this same level of skepticism is also levied upon ordinary scientific claims in the professional research journals. The standards are the same. Look what happened when the Utah chemists B. Stanley Pons and Martin Fleischmann claimed in a press conference to create "cold" nuclear fusion on their laboratory table. Scientists acted swiftly and skeptically. Within days of the announcement it became clear that no one could replicate the cold fusion results that Pons and Fleischmann claimed for their experiment. Their work was summarily dismissed. Similar plot lines unfold almost daily (minus the press conferences) for nearly every new scientific claim.

With scientists exhibiting such strong levels of skepticism, some people may be surprised to learn we heap our largest rewards and praises upon colleagues who succeed in discovering flaws in accepted paradigms. These same rewards also go to those who create new ways to understand the universe. Nearly all famous scientists, pick your favorite one, have been so praised in their own lifetimes. This path to success in one's professional career is antithetical to almost every other human establishment—especially to religion.

None of this is to say that the world does not contain religious scientists. In a recent survey of religious beliefs among math and science professionals,* 65 percent of the mathematicians (the highest rate) declared themselves to be religious, as did 22 percent

*Edward J. Larson and Larry Witham, *Nature* 394 (April 3, 1997): 313.

of the physicists and astronomers (the lowest rate). The national average among all scientists was around 40 percent and has remained largely unchanged over the past century. For reference, 90 percent of the American public claims to be religious (among the highest in Western society), so either nonreligious people are drawn to science or studying science makes you less religious.

But what of those scientists who are religious? One thing is for sure, successful researchers do not get their science from religious doctrines. But the methods of science have little or nothing to contribute to ethics, inspiration, morals, beauty, love, hate, or social mores. These are vital elements to civilized life, about which God in nearly every religion has much to say. What it all means is that for many scientists there is no conflict of interest.

When scientists do talk about God, they typically invoke him at the boundaries of knowledge where we are most humble and where our sense of wonder is greatest. Examples of this abound. During an era when planetary motions were on the frontier of natural philosophy, Ptolemy couldn't help feeling a religious sense of majesty when he penned, "When I trace at my pleasure the windings to and fro of the heavenly bodies, I no longer touch the earth with my feet. I stand in the presence of Zeus himself and take my fill of ambrosia." Note that Ptolemy was not weepy about the fact that the element mercury is liquid at room temperature, or that a dropped rock falls straight to the ground. While he could not have fully understood these phenomena either, they were not seen at the time to be on the frontiers of science and worthy of a religious epithet.

In the thirteenth century, Alfonso the Wise (Alfonso X), the king of Castile and León who also happened to be an accomplished academician, was frustrated by the complexity of Ptolemy's epicycles. Being less humble than Ptolemy, Alfonso once mused, "Had I been around at the creation, I would have given some useful hints for the better ordering of the universe."

In his 1687 masterpiece, *The Mathematical Principles of Natural Philosophy*, Isaac Newton lamented that his new equations of gravity, which describe the force of attraction between pairs of objects, would not maintain a stable system of orbits for multiple planets. Under this instability, planets would either crash into the Sun or be ejected from the solar system altogether. Worried about the long-term fate of Earth and other planets, Newton invoked the hand of God as a restoring force that would maintain a long-lived solar system. Over a century later, the French mathematician and dynamicist Pierre-Simon Laplace invented perturbation theory for his five-volume treatise *Celestial Mechanics* (1799–1825), which extended the applicability of Newton's equations to complex systems of planets such as ours. Laplace showed that our solar system was indeed stable and did not require the hand of a deity after all. When queried by Napoleon Bonaparte on the absence of any reference to an "author of the universe" in his book, Laplace replied, "I have no need of that hypothesis."

And in full agreement with King Alfonso's frustrations with the universe, Albert Einstein noted in a letter to a colleague, "If God created the world, his primary worry was certainly not to make its understanding easy for us." When Einstein could not figure out how or why a deterministic universe would require the roulette formalisms of quantum mechanics, he mused, "It is hard to sneak a look at God's cards. But that he would choose to play dice with the world . . . is something that I cannot believe for a single moment." When an experimental result was shown to Einstein that would have disproved his new theory of gravity, Einstein commented, "The Lord is subtle, but malicious he is not." The Danish physicist Niels Bohr, a contemporary of Einstein, heard one too many of Einstein's God remarks and declared that Einstein should stop telling God what to do!

Today, you hear the occasional astrophysicist (one in a fifty or so) invoke God when asked where did all our laws of physics come from, or what was around before the big bang. As we have come to anticipate, these questions comprise the modern frontier of cosmic discovery and, at the moment (like the above examples in their day), they transcend the answers our available data can supply.

Some promising ideas already exist that address these questions, such as inflationary cosmology and string theory. They may ultimately provide the answers, pushing back the boundary of our awe of the cosmos.

My personal views are entirely pragmatic, and partly resonate with those of Galileo who, during his trial, is credited with saying, "The Bible tells you how to go to heaven, not how the heavens go." Galileo further noted, in a 1615 letter to Madame Christina of Lorraine, the Grand Duchess of Tuscany, "In my mind God wrote two books. The first book is the Bible, where humans can find the answers to their questions on values and morals. The second book of God is the book of nature, which allows humans to use observation and experiment to answer our own questions about the universe."

I simply go with what works. And what works is the healthy skepticism embodied in the scientific method. Believe me, if the Bible had ever been shown to be a rich source of scientific answers and enlightenment, we would be mining it daily for cosmic discovery. Yet my vocabulary of scientific inspiration strongly overlaps with that of religious enthusiasts. I, like Ptolemy, am humbled in the presence of our clockwork universe. When I am on the cosmic frontier, and I touch the laws of physics with my pen, or when I look upon the endless sky from an observatory on a mountaintop, I well up with an admiration for its splendor. But I do so knowing and accepting that if I propose a God beyond that horizon, one who graces the valley of our collective ignorance, then the day will come again when our sphere of knowledge has grown so large that I will have "no need of that hypothesis."

One of the strongest footholds that religion retains on life is in death. The Bible and other doctrines of revealed truths have much to say about the afterlife. But once again I invoke neither assumption nor hypothesis. In the cycle of life, the total of all matter and energy remains unchanged—a fundamental feature of physical laws that falls closest to my heart. I have even taken the concept to a level of which the new-age movement would be proud. When I die, I want to be buried, not cremated. Whenever you burn organic matter, including human corpses, the chemical energy content of the body's quadrillion cells converts entirely into heat energy, which raises the atmospheric temperature near the crematorium, and eventually radiates back into space. When deposited into the universe, this low-energy thermal radiation increases cosmic entropy and is largely unrecoverable to perform any other further work.

I owe Earth (and the universe) much more than this. For my entire omnivorous life I have eaten of its flora and feasted on its fauna. Countless plants and animals have sacrificed their lives and unwillingly donated their energy content to my sustenance. The least I can do is donate my body back to this third rock from the Sun. I want to be buried, just like in the old days, where I decompose by the action of microorganisms, and I am dined upon by any form of creeping animal or root system that sees fit to do so. I would become their food, just as they had been food for me. I will have recycled back to the universe at least some of the energy that I have taken from it. And in so doing, at the conclusion of my scientific adventures, I will have come closer to the heavens than to Earth.

INDEX